SpringerBriefs in Electrical and Computer Engineering

Speech Technology

Series Editor

Amy Neustein

For further volumes:
http://www.springer.com/series/10059

Editor's Note

The authors of this series have been hand selected. They comprise some of the most outstanding scientists—drawn from academia and private industry—whose research is marked by its novelty, applicability, and practicality in providing broad-based speech solutions. The Springer Briefs in Speech Technology series provides the latest findings in speech technology gleaned from comprehensive literature reviews and *empirical investigations* that are performed in both laboratory and *real life* settings. Some of the topics covered in this series include the presentation of real life commercial deployment of spoken dialog systems, contemporary methods of speech parameterization, developments in information security for automated speech, forensic speaker recognition, use of sophisticated speech analytics in call centers, and an exploration of new methods of soft computing for improving human-computer interaction. Those in academia, the private sector, the self service industry, law enforcement, and government intelligence are among the principal audience for this series, which is designed to serve as an important and essential reference guide for speech developers, system designers, speech engineers, linguists, and others. In particular, a major audience of readers will consist of researchers and technical experts in the automated call center industry where speech processing is a key component to the functioning of customer care contact centers.

Amy Neustein, Ph.D., serves as editor in chief of the *International Journal of Speech Technology* (Springer). She edited the recently published book *Advances in Speech Recognition: Mobile Environments, Call Centers and Clinics* (Springer 2010), and serves as quest columnist on speech processing for Womensenews. Dr. Neustein is the founder and CEO of Linguistic Technology Systems, a NJ-based think tank for intelligent design of advanced natural language-based emotion detection software to improve human response in monitoring recorded conversations of terror suspects and helpline calls.

Dr. Neustein's work appears in the peer review literature and in industry and mass media publications. Her academic books, which cover a range of political, social, and legal topics, have been cited in the Chronicles of Higher Education and have won her a pro Humanitate Literary Award. She serves on the visiting faculty of the National Judicial College and as a plenary speaker at conferences in artificial intelligence and computing. Dr. Neustein is a member of MIR (machine intelligence research) Labs, which does advanced work in computer technology to assist underdeveloped countries in improving their ability to cope with famine, disease/ illness, and political and social affliction. She is a founding member of the New York City Speech Processing Consortium, a newly formed group of NY-based companies, publishing houses, and researchers dedicated to advancing speech technology research and development.

K. Sreenivasa Rao

Predicting Prosody from Text for Text-to-Speech Synthesis

 Springer

K. Sreenivasa Rao
School of Information Technology
Indian Institute of Technology Kharagpur
Kharagpur
West Bengal, India

ISSN 2191-8112   ISSN 2191-8120 (electronic)
ISBN 978-1-4614-1337-0 ISBN 978-1-4614-1338-7 (eBook)
DOI 10.1007/978-1-4614-1338-7
Springer New York Heidelberg Dordrecht London

Library of Congress Control Number: 2012936122

Printed on acid-free paper

Springer is part of Springer Science+Business Media (www.springer.com)

Preface

During production of speech human beings impose durational constraints and into-nation patterns on the sequence of sound units to convey the intended message. This inherent ability of the human beings in using the prosody (duration and intonation) knowledge is naturally acquired, and is difficult to articulate. But for synthesizing speech from a text by a machine, it is necessary to acquire, represent and incorporate this prosody knowledge. The prosody constraints are not only characteristic of the speech message and the language, but they also characterize a speaker uniquely. Even for speech recognition, human beings seem to rely on the prosody cues to disambiguate errors in the perceived sounds. Thus acquisition and incorporation of prosody knowledge becomes important for developing speech systems. This book attempts to discuss the methods to capture the prosody knowledge in speech in Indian languages, and to incorporate the knowledge in speech systems.

The book presents an approach to capture the implicit duration and intonation knowledge using models of neural networks and support vector machines. The results are demonstrated using labeled database for speech in three Indian languages, namely, Hindi, Telugu and Tamil. The prosody models are shown to possess speaker and language characteristics as well, besides information about the message in speech. Thus the models can be explored for identification of speaker and language. An important application of prosody models is in the synthesis of speech from text, and in voice conversion. For this, the prosody knowledge captured in the models needs to be incorporated in the speech signal. In this book a flexible prosody modi-fication method is discussed. The method is based on exploiting the nature of exci-tation of speech production mechanism. The book also discuss methods to modify the speech prosody in real time, modification of formant structure and imposing the desired pitch contour.

This book mainly intended for speech researchers working on prosodic aspects of speech such as characterization, representation, acquisition and incorporation of speech prosody. This book is also useful for the young researchers, who want to pursue the research in speech processing. This book can be kept as a text book or reference book for the postgraduate level advanced speech processing course. The book has been organized as follows:

Chapter 1 introduces the prosodic aspects of speech in view of its significance for developing various speech systems and its manifestation in speech for providing the naturalness to speech. Chapter 2 provides the review of duration and intonation models for acquisition of duration and intonation knowledge. Methods for modification of prosody of speech are also reviewed in this chapter. Chapter 3 presents the analysis of durations of sound units with respect to positional and contextual factors. Chapters 4 and 5 discuss modeling of duration and intonation patterns of syllables using Feedforward Neural Network (FFNN) and Support Vector Machines (SVMs). Chapter 6 discusses on the incorporation of duration and intonation information in a speech signal. Chapter 7 illustrates some important practical applications based on prosody modification. Chapter 8 summarizes the contents of the book, highlights the contributions of the chapters and discusses the scope for future work.

Many people have helped me during the course of preparation of this book. I would especially like to thank Professor B. Yegnanarayana for the constant encouragement and technical discussions during course of editing and organization of the book. Special thanks to my colleagues at IIT Madras during my PhD work for their cooperation and coordination to carry out the work. I would like to thank my faculty colleagues and administration staff at IIT Guwahati and IIT Kharagpur for helping me during editing and compilation time. I am grateful to my parents and family members for their constant support and encouragement. Finally, I thank all my friends and well-wishers.

Kharagpur, West Bengal, India   K. Sreenivasa Rao

Contents

Acronyms

AANN	Autoassociative Neural Network
CART	Classification and Regression Tree
CV	Consonant Vowel
DCT	Discrete Cosine Transform
DFT	Discrete Fourier Transform
FD-PSOLA	Frequency Domain Pitch Synchronous Overlap and Add
F	Female
FFNN	Feedforward Neural Network
GC	Glottal Closure
Hi	Hindi
HNM	Harmonic plus Noise Model
IDFT	Inverse Discrete Fourier Transform
IITM	Indian Institute of Technology Madras
IPCG	Inter Perceptual Center Group
IPO	Institute of Perception Research
ITRANS	Common transliteration code for Indian languages
LP	Linear Prediction
LPCs	Linear Prediction Coefficients
LP-PSOLA	Linear Prediction Pitch Synchronous Overlap and Add
M	Male
MOS	Mean Opinion Score
NN	Neural Network
OLA	Overlap and Add
PF	Phrase Final
PI	Phrase Initial
PM	Phrase Middle
PSOLA	Pitch Synchronous Overlap and Add
RFC	*rise/fall/connection*
RNN	Recurrent Neural Network
SOP	Sums-of-Product

STRAIGHT	Speech Transformation and Representation using Adaptive Interpolation of weiGHTed spectrum
SVM	Support Vector Machine
Ta	Tamil
TD-PSOLA	Time Domain Pitch Synchronous Overlap and Add
Te	Telugu
TTS	Text-to-Speech
VOP	Vowel Onset Point
V/UV/S	Voiced/Unvoiced/Silenc
WF	Word Final
WI	Word Initial
WM	Word Middle

Chapter 1
INTRODUCTION

Abstract This chapter discuss about the significance of prosody knowledge for developing speech systems by machine, and for performing various speech tasks by human beings. The manifestation of prosodic knowledge in speech at linguistic, articulatory, acoustic and perception levels is described. Some of the inherent prosodic knowledge sources present in speech, which can be analyzed by machine automatically are also discussed in this chapter. The basic objective, scope and organization of the contents of the book are discussed at the end of this chapter.

1.1 Significance of Prosody Knowledge

Human beings use durational and intonation patterns on the sequence of sound units, while producing speech. It is these prosody constraints (duration and intonation), that lend naturalness to human speech. Lack of this knowledge can easily be perceived, for instance, in the speech synthesized by a machine. Even though human beings are endowed with this knowledge, they are not able to express it explicitly. But it is necessary to acquire, represent and incorporate this prosody knowledge for synthesizing speech from a text. Speech signal carries information about the message to be conveyed, speaker and language in the prosody constraints, and these prosody cues aid human beings to get the message, and identify speaker and language. The prosody knowledge also helps to overcome perceptual ambiguities. Thus, acquisition and incorporation of prosody knowledge is essential for developing speech systems.

K.S. Rao, *Predicting Prosody from Text for Text-to-Speech Synthesis*, SpringerBriefs
in Electrical and Computer Engineering, DOI 10.1007/978-1-4614-1338-7_1,
© Springer Science+Business Media New York 2012

1.2 Manifestation of Prosody Knowledge

Prosody can be viewed as speech features associated with larger units (than phonemes) such as syllables, words, phrases and sentences. Consequently, prosody is often considered as suprasegmental information. The prosody appears to structure the flow of speech, and is perceived as melody and rhythm. The prosody is represented acoustically by a pattern of duration and intonation (F_0 contour). The prosody can be distinguished at four principal levels of manifestation [1]. They are at (a) Linguistic intention level, (b) articulatory level, (c) acoustic realization level and (d) perceptual level.

At the linguistic level, prosody refers to relating different linguistic elements to each other, by accentuating certain elements of a text. For example, the linguistic distinctions that can be communicated through prosodic means are the distinction between question and statement, or the semantic emphasis of an element with respect to previously enunciated material.

At the articulatory level, the prosody is physically manifested as a series of articulatory movements. Thus, prosody manifestations typically include variations in the amplitudes of articulatory movements, variations in air pressure, and specific patterns of electric impulses in nerves leading to the articulatory musculature.

Muscle activity in the respiration system, and along the vocal tract leads to emission of sound waves. The acoustic realization of prosody can be observed and quantified using acoustic signal analysis. The main acoustic parameters bearing on prosody are fundamental frequency (F_0), intensity and duration. For example, stressed syllables have higher fundamental frequency, greater amplitude and longer duration than unstressed syllables.

Finally, speech sound waves enter the ear of the listener who derives the linguistic and paralinguistic information from prosody via perceptual processing. At the level of perception, prosody can be expressed in terms of subjective experience of the listener, such as pauses, length, melody and loudness.

It is difficult to process or analyze the prosody through speech production or perception mechanisms. Hence the acoustic properties of speech are exploited for analyzing the prosody. In the next section we will discuss some of the sources of knowledge that are present in the speech signal.

1.3 Implicit Knowledge in Speech Signal

For illustration, some of the knowledge sources present in speech signal are demonstrated here. Fig. 1.1 shows a speech signal and its transcription, energy contour, pitch contour and spectrogram. The waveform shown in Fig. 1.1(a) represents the time domain representation of a speech signal, the abscissa (X-axis) indicates the timing information and the ordinate (Y-axis) indicates the amplitude of speech samples. The transcription (Fig. 1.1(b)) represents the sequence of sound units and their boundaries. This gives information about the identities of sound units present in the

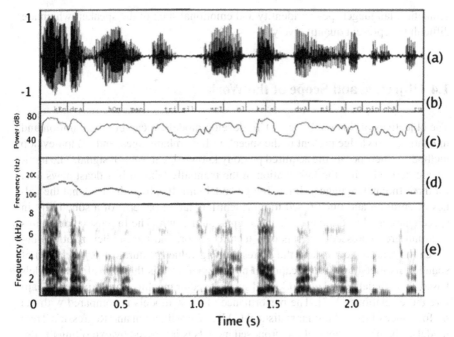

Fig. 1.1 (a) Speech signal, (b) Transcription for the utterance *"kEndra hOm mantri srI el ke advAni ArOpinchAru"*, (c) Energy contour, (d) Pitch contour and (e) Wideband spectrogram.

speech signal and their durations. Energy contour (Fig. 1.1(c)) indicates the distribution of energy in different regions of the speech signal, and also gives a rough indication of the voiced and nonvoiced regions. Pitch contour (Fig. 1.1(d)) indicates the global and local patterns of intonation. Global intonation pattern refers to the characteristics of the whole sentence or phrase. A rising intonation pattern at the global level indicates that the sentence (phrase) is interrogative, and a declining intonation pattern indicates a declarative sentence. Local fall-rise patterns indicate the nature of words and basic sound units. The spectrogram (Fig. 1.1(e)) is used to represent the speech intensity in different frequency bands as a function of time. In Fig. 1.1(e), the ordinate is the frequency axis, and the grey value indicates the energy (intensity) of speech signal. The dark bands in the spectrogram represent the resonances of the vocal tract system. These resonances are also called formant frequencies, which represent the high energy regions in the frequency spectrum of a speech signal. These formant frequencies are distinct for each sound unit. The shape of the sequence of dark bands indicates the changes in the shape of the vocal tract from one sound unit to the other. Speech signal also contains information about semantics, language, speaker identity and emotional state of the speaker, which are difficult to represent quantitatively.

1.4 Objective and Scope of the Work

The objectives of this work are: (1) To develop models to capture the duration and intonation knowledge present in the speech in Indian languages, and (2) to develop methods to incorporate the acquired prosody knowledge in speech signal. The models are derived using the information in the manually labeled broadcast news data for three Indian languages, Hindi, Telugu and Tamil. It is hypothesized that the values of duration and pitch (used to represent intonation pattern) of a sound unit in speech is related to the linguistic constraints on the unit. The linguistic constraints of a unit are expressed in terms of a feature vector, and the implicit relations between the feature vectors and the corresponding values of duration or pitch (F_0) of sound units are sought to be captured by a model. In this thesis nonlinear models based on artificial neural networks and support vector machines are explored to capture these relations [2, 3]. The performance of these models is compared with the performance of more familiar statistical CART (Classification and Regression Tree) models. The performance of the proposed models is improved by exploiting the dependency of the duration and intonation on each other. Some postprocessing of the output of the models, and preprocessing the input by dividing it into different categories, help further improvement in the performance of the models. The ability of the prosody models in capturing speaker-specific and language-specific information is demonstrated by discriminating speakers and languages using these models.

To incorporate the acquired prosody knowledge into a speech signal, a new method for prosody modification of speech is proposed. The method involves deriving a modified excitation signal such as Linear Prediction (LP) residual. The modification of the excitation signal (according to the desired prosody) is based on the knowledge of the instants of significant excitation of the vocal tract system during production of speech [4]. A computationally efficient method is proposed to determine the instants of significant excitation from speech, using the Hilbert envelope and group delay function of the LP residual. A method for duration modification using vowel onset points and instants of significant excitation is proposed, which provides flexibility for modification in different regions. Methods are developed for imposing the pitch contour on a given speech signal, and for modification of formant structure according to pitch variations.

1.5 Organization of the Book

The evolution of ideas presented in this book is given briefly in Table 1.1. The book is organized as follows:

A review of duration and intonation models for acquisition of duration and intonation knowledge is presented in Chapter 2. Methods for modification of prosody of speech are also reviewed in this chapter, and motivation for the present work is laid.

In Chapter 3, the analysis of durations of sound units is presented. Durations of the syllables are analyzed with respect to positional and contextual factors. For detailed duration analysis, syllables are categorized into groups based on size of the word and position of the word in the utterance, and the analysis is performed separately on each category.

Chapter 4 presents modeling of durations of syllables. Feedforward Neural Network (FFNN) and Support Vector Machines (SVMs) are proposed for predicting the durations of syllables. The effect of different constraints and their interactions in modeling the durations of syllables is examined. Finally, the proposed duration models have demonstrated for capturing the language-specific and speaker-specific information.

Chapter 5 discuss about modeling of intonation patterns using FFNNs and SVMs. Postprocessing methods using intonation constraints are proposed for improving the accuracy of prediction. The ability of the intonation models to capture the language-specific and speaker-specific information is demonstrated.

Chapter 6 focuses on the incorporation of duration and intonation information in a speech signal. A new method is proposed for the modification of pitch and duration of a given speech signal using the instants of significant excitation as anchor points. The performance of the proposed method for prosody modification is compared with some standard methods.

Chapter 7 discuss about the practical aspects of prosody modification. A computationally efficient method for determining the instants of significant excitation is proposed. The application of proposed duration and intonation prediction models has been demonstrated using concatenative text to speech synthesis system. A new duration modification method using vowel onset points and instants of significant excitation is proposed. For voice conversion application, methods are developed to modify the formant structure and to impose the desired pitch contour on a given speech signal.

Chapter 8 summarizes the research work presented in this book, highlights the contributions of the work and discusses the scope for future work.

Table 1.1 Evolution of ideas presented in the book.

**Acquisition and incorporation of prosody knowledge
for speech systems in Indian languages.**

- Human beings are endowed with prosody knowledge and use it for various speech related tasks (speech production, perception and identifying the speaker and language).

- For developing speech systems, acquisition and incorporation of prosody knowledge (duration and intonation) is essential.

- Hypothesis: The values of duration and pitch (used to represent intonation pattern) of a sound unit in speech is related to the linguistic constraints on the unit.

- Acquisition of prosody knowledge using rule-based (manual) methods is difficult.

- Nonlinear models are known to capture the implicit relation between the input and output. Hence nonlinear models based on artificial neural networks and support vector machines are explored to capture the duration and intonation knowledge present in speech.

    - In speech signal, duration and intonation patterns are interrelated at higher level. Performance of the prosody models is improved by exploiting the dependency of the duration and intonation on each other.
    - Preprocessing and postprocessing methods, help further improvement in the performance of the models.

- The prosody models are shown to capture speaker and language characteristics as well, besides information about the message in speech. Hence these models can be explored for identification of speaker and language.

- The acquired prosody knowledge needs to be incorporated in the speech signal, for text-to-speech synthesis and voice conversion applications.

    - A method is proposed for modification of prosody of speech using instants of significant excitation.
    - A computationally efficient method is developed for online implementation of prosody modification of speech.
    - For voice conversion application, methods are developed to modify the formant structure and to impose the desired pitch contour on a given speech signal.
    - A flexible duration modification method is developed using vowel onset points and instants of significant excitation.

Chapter 2
PROSODY KNOWLEDGE FOR SPEECH SYSTEMS: A REVIEW

Abstract This chapter provides an overview of some of the existing models for acquisition of prosody and also methods for incorporating the prosody in speech signal. The models used for capturing the prosody range from rule-based models to data-based models. Rule-based models often generalize too much and can not handle exceptions well. Data-based models are generally dependent on the quality and quantity of the data available. At the end of the chapter motivation for the present work in this book is derived from the limitations of the existing works.

2.1 Models for Acquisition of Duration Knowledge

The duration models are generally grouped into rule-based models and statistical models [5]. The main difference between rule-based and statistical models is that a rule-based model can be built on relatively less speech data. The formation of the rules, however, requires expert knowledge and considerable optimization effort by trial and error. In contrast, statistical models are built from a large amount of labeled speech data. In this section we present five duration models. Among them, Klatt duration model and IITM duration model belongs to rule-based models, and Sums-of-Products (SOP) model, Classification and Regression Trees (CART) and Neural Network (NN) models are statistical models.

2.1.1 Klatt duration model

Dennis Klatt proposed a rule-based model [6], which was implemented in the MITalk system [7]. The model was based on information presented in linguistic and phonetic literature about different factors affecting segmental durations. The

K.S. Rao, *Predicting Prosody from Text for Text-to-Speech Synthesis*, SpringerBriefs in Electrical and Computer Engineering, DOI 10.1007/978-1-4614-1338-7_2,
© Springer Science+Business Media New York 2012

duration (d) of each segment (phone) was calculated according to the following relation:

$$d = (d_i - d_m) * \tfrac{P}{100} + d_m$$

where d_i and d_m are the inherent and minimum durations for the phone, respectively. This model had 10 rules that were based on the effects of the phonetic environment, emphasis and stress level on the duration of the current phone. Each rule adjusts the P term in a multiplicative manner, and the final result is the sum of the effect of final rule, and the product of the effects of the rest of the rules [7]. Like other rule-based models, the rules and their parameter values in Klatt model are determined manually and by trial and error. Following the Klatt model, similar models were developed for other languages like German and French [8, 9].

2.1.2 IITM duration model

S. R. Rajesh Kumar *et al* [10–12] proposed a rule-based duration model for developing a text-to-speech system for the Indian language Hindi. The TTS system is based on parameter concatenation approach. The basic units in the system are the characters of the Indian language Hindi. The units were collected in a neutral context using carrier words. The rules were derived by analyzing 500 sentences spoken by a native male speaker. The analysis was performed with respect to positional and contextual factors. The final rule-base consists of 31 rules, represented in the form of IF-THEN rules. The format of a rule is as follows:

> *IF* < *number of antecedents* >
> *antecedent* 1
> *antecedent* 2
>   .
>   .
>
> *THEN* < *number of consequents* >
> *consequent* 1
> *consequent* 2
>   .
>   .

For example, a rule could be as follows:

> *IF* 2
> *unit belongs to CV category*
> *next character is fricative* /*ha*/
> *THEN* 1
> *decrease duration by* 25%

The above rule states that if the present unit is of CV category, and the next character is /ha/, then decrease the base duration of the present character by 25%. The rules are activated by means of rule-based inference engine (or a rule interpreter). The rules combine multiplicatively, if more than one rule fires for the same unit. Later the rule-base was upgraded by analyzing large amount of broadcast news data for the language Hindi [13]. Parallelly, a separate rule-base was developed for the language Telugu by analyzing the broadcast news data [14].

In general, rule-based models have the following disadvantages [15]: (1) Manually exploring the effect of mutual interactions among linguistic features at different levels is difficult. (2) The rule-inference process usually involves a controlled experiment, in which only a limited number of contextual factors are examined. The resulting inferred rules may therefore not be general enough for unlimited texts.

2.1.3 Sums-of-products model

Van Santen proposed a sums-of-products model, which is a statistical model [16, 17]. Also expert knowledge was used concerning different factors and their interactions influencing the segment durations [16–21]. The model is linear, and is based on prior phonetic and phonological information as well as information collected by analyzing data. Segment durations $d(.)$ are calculated as a function of a feature vector x by the following equation:

$$d(x) = \sum_{i=1}^{K} \prod_{j=1}^{I_i} S_{i,j}(d_j)$$

where K is number of product terms in the model, I_i is the set of indices of factors included in the ith product term and $S_{i,j}(.)$ indicates factors at different levels. For example, the duration of a vowel /e/ which is followed by a voiced consonant and is in utterance final position is given by taking the intrinsic duration of the vowel $\alpha(/e/)$, adding the effect of being utterance final position $\delta(Final)$, and finally adding the effect of post-vocalic voicing $\beta(Voiced)$ modulated by utterance-finality $\gamma(Final)$. This is represented by the following equation:

$d(Vowel : /e/, Next : Voiced, Loc : Final) = \alpha(/e/) + \delta(Final) + \beta(Voiced) * \gamma(Final)$

The model ignores the direction of influence of various factors on the duration of the unit. Once the structure of an SOP model is developed in terms of the input factors and their interactions, the parameter estimation for the model is obtained using least-squares method [18]. The advantage of the SOP model is that the model can be developed using small amount of data. But in the SOP model, the number of different sums-of-products grows exponentially with the number of factors. In addition, the SOP model requires significant preprocessing of the data to correct the interactions among factors and data imbalance [22, 23].

2.1.4 Classification and regression tree model

The Classification and Regression Trees (CART) are typical data-based duration models that can be constructed automatically. Their self-configuration feature makes them popular [24]. In the CART model a binary branching tree is constructed by feeding the attributes of the feature vectors from top node, and passing the attributes through the arcs representing the constraints [25]. Fig. 2.1 shows a partial tree constructed to predict the durations of syllables for the Indian language Telugu. The

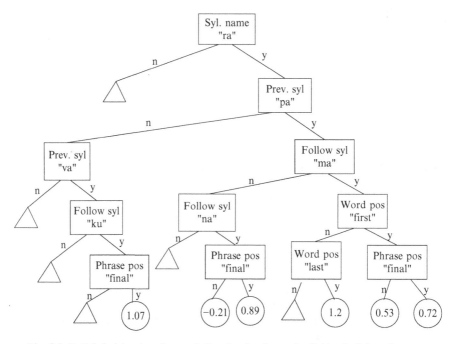

Fig. 2.1 Partial decision tree for predicting the durations of syllables in Telugu language (syl: syllable, prev: previous, pos: position, y: yes and n: no). The triangles indicate omitted sections.

numbers in the leaf nodes are called *z-scores*, and the final durations are calculated according to the equation $d = \mu + z_s * \sigma$, where d is duration of a syllable, μ is the mean duration, z_s is *z-score* value, and σ is the standard deviation of the syllable duration. Both the mean and standard deviation are estimated from a corpus. The triangles in the figure shows the sections omitted. The tree assigns different durations for the syllable /ra/ when it occurs in different contexts. A duration value of 178.4 ms ($128 + 42 * 1.2 = 178.4$, where 128 and 42 are mean and standard deviation, respectively, of the syllable /ra/ in the database) is assigned when it satisfies the following criteria: Previous and following syllables are /pa/ and /ma/, respectively, and the syllable is in last word of the phrase.

The algorithm clusters durations according to their contexts [25, 26]. The contextual factors include the stress level of the current phone, its position in the word, its position in the phrase, and the phonetic context on either side of the current one. Using individual phone labels to describe the phonetic context would result in a serious data scarcity problem. Therefore, it is better to describe the context in terms of broad classes, i.e., the phones can be grouped according to their phonological features. The tree constructing algorithms usually guarantee that the tree fits the training data well, but there is no guarantee that new and unseen data will be properly modeled. Sometimes the tree may be over-trained to fit the peculiarities in the training data [27]. This problem can be avoided by pruning the tree by cutting off the branches that are responsible for the over-training. The pruning is usually done by evaluating the tree on some unseen data (usually called the *pruning set*), and then simplifying the tree by cutting off the parts that do not fit well to the pruning data. CART models were developed for the languages Czech, Korean, Hindi and Telugu [28–30]. The prediction performance of the CART model depends on the coverage of the training data. They underperform when large percentage of data is missing.

2.1.5 Neural network model

Campbell [31, 32] has devised a neural network model which predicts the durations of syllables, and then derives phone durations from syllable durations. The networks are expected to learn the underlying interactions in the context. That is, they can represent the rule governing the behavior implicit in the data.

Feature vector for each syllable consists of information about the number of phonemes in a syllable, the nature of syllable, position in the tone group, type of foot, stress and word class. Syllable durations are predicted with these feature vectors as input to the neural network. The durations of the phones are estimated from the predicted durations of the syllables using the *elasticity principle*. Under this principle, duration of each phone is approximately proportional to the probability density of its duration. The duration of each phoneme in the syllable is computed by solving the following equation for k:

$$d = \sum_{i=1}^{n} exp(\mu_i + k\sigma_i)$$

where d is the predicted duration of the syllable, n is number of phonemes (segments) in the syllable, μ_i and σ_i are mean and standard deviation of the log-transformed duration for the phoneme i, and k is a constant common for all phonemes within a syllable, which satisfies the above equation. For the phone i, the assigned duration will be $exp(\mu_i + k\sigma_i)$.

Barbosa and Bailly also used a neural network model to capture the perception of rhythm in speech [33]. The model predicts the duration of a unit, known as Inter-Perceptual Center Group (IPCG). The IPCG is delimited by the onset of a nuclear vowel and the onset of the following vowel. The model is trained with the following information: Boundary marker at phrase level, sentence mode, accent marker, number of consonants in IPCG, number of consonants in coda and nature of the vowel. Similar to the Campbell model, this model also proceeds in two steps. Firstly, the duration of the IPCG is computed by a sequential neural network. In the second step the IPCG duration is distributed among its segments using the elasticity principle. This model can deal with different speech rates and pauses [34]. Other neural network based models were also studied for Spanish, Arabic, German and Portuguese [35–38].

Some of the drawbacks in using neural network models are the following: (1) It is difficult to code (quantify) qualitatively different features. (2) Only fixed-form data types and structures are explicitly supported. (3) Usage of control structures is difficult. (4) Designing of optimal network configuration and interpretation of the interactions of the linguistic constraints is difficult.

2.2 Models for Acquisition of Intonation Knowledge

The fundamental frequency (F_0) of a human speech utterance is determined by a combination of many factors, from several levels of human speech production process [39]. At the lowest level, the local segmental factors that affect F_0 (micro-intonation) are caused by the dynamics of the human speech production process. At a higher level, F_0 contour is affected by the patterns of stress, melody and rhythm. Finally, the F_0 contour is also affected by gender, attitude, physical and emotional state of the speaker. In the past 20 years two major classes of intonation models have been developed. They are *phonological* models and *acoustic-phonetic* models [40]. Phonological models (tone sequence models) represent the prosody of an utterance as a sequence of abstract units. F_0 contour is generated from a sequence of phonologically distinctive tones, which are locally determined and do not interact with each other. Acoustic-phonetic models (superposition or overlay models) interpret F_0 contour as the result of the superposition of several components of different temporal scopes. Apart from these two prevalent types of intonation models, there are several important models that defy categorization as being either the superpositional or the tone sequence type. For instance, *perception-based* models exploit some acoustic-prosodic properties of speech, which are measurable and not perceived by the listener all the time. These models tend to downplay the linguistic functions of the intonational events. Finally, *acoustic stylization* models aim at efficient analysis and synthesis of F_0 contours using *rise/fall/connection* (RFC) model [41, 42].

2.2.1 Phonological model (Tone sequence model)

Tone sequence model was initially developed for English by Pierrehumbert [43]. In this model an utterance consists of *intonational phrases*, which are represented in terms of sequences of *tones* H and L, for high and low tone, respectively. These tones are in phonological opposition. In addition to tones, the model incorporates *accents* of three different types: *pitch accents, phrase accents* and *boundary tones*. Pitch accents are marked by '*' in the superscript, like H^* or L^*. Pitch accents may consists of two elements, e.g., L^*H. Phrase accents are marked by using "−", like $H−$. Boundary tones are marked by using "%". Phrase accents are used to mark pitch movements between pitch accents and boundary tones. Boundary tones are used at the edges (boundaries) of intonational phrases.

The occurrence of three accent types are constrained by a grammar, which can be described by a finite-state automaton. The grammar will generate or accept only well-formed intonational representations. The grammar for describing intonation contours or tunes for English can be formulated in the following regular expression. It states that an intonation phrase consists of three parts: one or more pitch accents, followed by phrase accent and ending with a boundary tone:

$$\left\{ \begin{array}{c} H^* \\ L^* \\ H^*+L \\ H+L^* \\ L^*+H \\ L+H^* \\ H^*+H \end{array} \right\} + \left\{ \begin{array}{c} H- \\ L- \end{array} \right\} \left\{ \begin{array}{c} H\% \\ L\% \end{array} \right\}$$

Sentences given in abstract tonal representation are converted to F_0 contours by means of *phonetic realization rules*. Tone sequence models have been implemented for the languages German, English, Chinese, Navajo and Japanese [44, 45]. Phonological models do not properly represent actual pitch variations. No distinction is made on the differences in tempo or acceleration of pitch movements. The temporal information is also not modeled. Phonological models are not easily ported from one language to another, since the inventory of categories must be thoroughly reviewed by expert linguists [46].

2.2.2 Acoustic-phonetic model (Fujisaki model)

Gronnum developed an intonation model for Danish language, which is conceptually quite different from the tone sequence model [47, 48]. The model is hierarchically organized and includes several simultaneous, non-categorical components of different temporal scopes. The components are *layered*, i.e., a component of short temporal scope is superimposed on a component of a longer scope. The intonation model developed by Lund also analyzes the intonation contour of an utterance as the result of the effects of several factors [49].

A classical *superpositional* intonational model has been presented by Fujisaki [50–52]. This model is widely used for the languages Japanese, German, English, Greek, Polish, Spanish and French [53–57]. The model is based on the assumption that any F_0 contour consists of two kinds of elements: the slowly varying phrase component and a quickly varying accent component. These components are said to be related to the actions of the laryngeal muscles, specifically the cricothyriod muscle, which control the frequency of vibration of the vocal chords. Thus the model has a physiological basis.

The model is driven by a set of commands in the form of impulses (the phrase commands) and stepwise functions (the accent commands), which are fed to a critically damped second order linear filter, and then superimposed to produce the final F_0 contour. The modeling process is shown in Fig. 2.2. While applying the Fujisaki

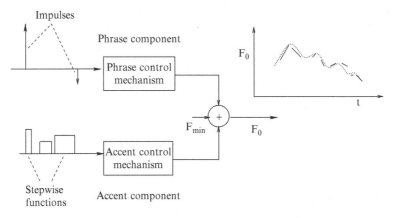

Fig. 2.2 Superpositional model (Fujisaki model) with phrase and accent components.

model to new languages, proper mapping is to be established between the prosody units and structures in the language to the model commands. One of the drawbacks of this model is that, it is difficult to extend the model to deal with low accents or slowly rising sections of contour [58].

2.2.3 Perception model (IPO model)

This model was developed at IPO (Institute of Perception Research, Eindhoven). In this model, certain F_0 movements are treated as perceptually relevant. Intonation analysis according to the IPO model consists of three steps [59]. Firstly, the perceptually relevant movements in F_0 contour are *stylized* by straight lines. This procedure results in a sequence of straight lines (known as *close copy contour*), that is perceptually indistinguishable from the original intonation contour. These two contours (original and *close copy*) are known to be *perceptually equivalent*. The motivation for stylizing the original intonation contour is that the variability of raw F_0 curves presents a serious obstacle for finding regularities. In the second step, an inventory of F_0 movements is developed by expressing the common features of the close copy contours in terms of duration and range of F_0. In the third and final step, a grammar of possible and permissible combination of F_0 movements is derived. Example of a stylized F_0 contour derived from the natural intonation pattern is shown in Fig. 2.3. Stylization in the IPO model is performed by a human

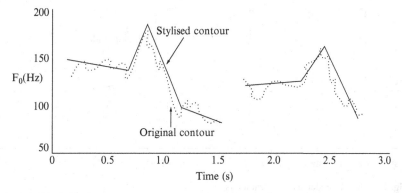

Fig. 2.3 Example of a stylized F_0 contour (solid line) derived from original intonation pattern (dotted curve).

experimenter. Methods for automatic stylization of intonation contours on perceptual grounds have been proposed by Mertens and d'Alessandro [60, 61]. The IPO model was originally developed for Dutch. Later it was applied to English, German and Russian [62–65]. There are two major drawbacks observed in the IPO model [58]. Firstly, the resynthesized contours derived from the straight line approximations are quite different from the original. This is due to difficulty in modeling a curve with straight lines. The second problem arises with the use of limited set of levels for approximating the raw F_0 contour.

2.2.4 Acoustic stylization model (Tilt model)

The tilt intonation model was designed to provide a robust computational analysis and synthesis of intonation contours. This model is based on *rise/fall/connection* (RFC) model, and is a bi-directional model that gives an abstraction directly from the data [41, 42]. The abstraction can then be used to produce a close copy of the original contour. In tilt model, each intonation event, be an *accent*, a *boundary, silence* or a *connection* between events, is described by a set of continuous parameters, which are useful for prosody control in speech synthesis. The events are described by the following parameters as shown in Fig. 2.4: Starting F_0, rise amplitude (A_r), fall amplitude (A_f), duration, peak position and the tilt (t). Tilt is defined as the ratio of the difference between the rise and fall amplitudes to the sum of the rise and fall amplitudes [41].

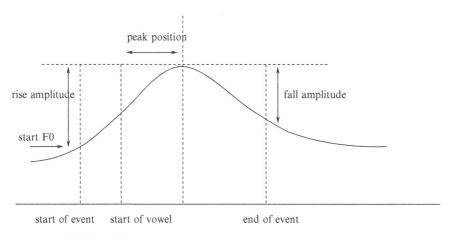

Fig. 2.4 The tilt model and its parameters.

$$t = \frac{|A_r| - |A_f|}{|A_r| + |A_f|}$$

The tilt parameter gives the actual shape of the event with a range from -1 to 1, where -1 indicates pure fall, 0 is a symmetric peak and 1 is pure rise. The shape in Fig. 2.4 has a tilt value of approximately 0.25.

The important aspects of the tilt model are: (1) Ability to capture both phonetic and phonological aspects of intonation, (2) ability to generate F_0 contours from the linguistic representation, and (3) derive the linguistic representation automatically from the utterance's acoustics. The tilt model gives a good representation of the natural F_0 contour and a better description of the type of accents, but it does not provide a linguistic interpretation for different types of accents.

2.2.5 Neural network model

Several models based on neural networks principles are described in the literature for predicting the intonation patterns of syllables in continuous speech [37, 46, 66–69]. Scordis and Gowdy used neural networks in parallel and distributed manner to predict the average F_0 value for each phoneme, and the temporal variations of F_0 within a phoneme [66]. The network consists of two levels: Macroscopic level and microscopic level. At the macroscopic level, a Feedforward Neural Network (FFNN) is used to predict the average F_0 value for each phoneme. The input to the FFNN consists of the sequence of phonemic symbols (identities of phonemes), which represents the contextual information. At the microphonemic level, a Recurrent Neural Network (RNN) is used to predict the temporal variations of F_0 within a phoneme. Marti Vainio and Toomas Altosaar used a three layer FFNN to predict F_0 values for the sequence of phonemes in Finnish language [67]. The features used for developing the models are: Phoneme class, length of a phoneme, identity of a phoneme, identities of previous and following syllables (context), length of a word and place of a word. Buhmann *et al* used a RNN for developing multi-lingual intonation models [46]. The features used in this work are the universal (language independent) linguistic features such as part-of-speech and type of punctuation, combined with prosody features such as word boundary, prominence of the word and duration of the phoneme.

2.2.6 IITM intonation model

An intonation model for the Indian language Hindi was proposed while developing a TTS system [70, 71]. The analysis was performed on reading style of speech. A corpus of 500 sentences was read out by two native male speakers. The analysis was performed at different levels to capture the following intonation patterns:

- Intonation patterns for simple declarative and interrogative sentences
- Local fall-rise patterns
- Resetting of F_0 contour
- Effect of segmental properties on F_0 contour

The captured knowledge is represented in the form of IF-THEN rules. The rules are classified into three groups. They are: (1) Pitch accent rules (macro prosody), (2) segmental prosodic variations (micro prosody) of F_0 contour, and (3) rules for pause insertion. For incorporating the intonation knowledge, the input text is parsed at sentence, word and character levels. At each level the associated intonation knowledge is incorporated by activating the corresponding rules. For example a rule can be as follows:

> *IF* 2
> *word is at sentence beginning*
> *word is disyllabic*
> *THEN* 2
> *select the middle of the second vowel*
> *fix the pitch accent value as* 180 *Hz*

2.3 Approaches for Incorporation of Prosody Knowledge

The acquired prosody knowledge needs to be incorporated in the speech signal for developing different speech systems. Here incorporation of prosody knowledge means modification of prosody parameters (duration and intonation) of speech. The objective of prosody modification is to alter the pitch contour and durations of the sound units of speech without affecting the short-time spectral envelopes [72]. Prosody modification is useful in a variety of applications related to speech communication [73–75]. For instance, in a text-to-speech system, it is necessary to modify the durations and pitch contours of the basic units and words in order to incorporate the relevant suprasegmental knowledge in the utterance corresponding to the sequence of these units [76]. Time-scale (duration) expansion is used to slow down rapid or degraded speech to increase the intelligibility [77]. Time-scale compression is used in message playback systems for fast scanning of the recorded messages [77]. Frequency-scale modification is often performed to transmit speech over limited bandwidth communication channels, or to place speech in a desired bandwidth as an aid to the hearing impaired [78]. While pitch-scale modification is useful for a TTS system, formant modification techniques are also used to compensate for the defects in the vocal tract and for voice conversion [74, 79, 80].

Several approaches are available in the literature for prosody modification [72, 75, 77, 81–94]. Approaches like Overlap and Add (OLA), Synchronous Overlap and Add (SOLA) and Pitch Synchronous Overlap and Add (PSOLA) operate directly on the waveform (time domain) to incorporate the desired prosody information [75, 83, 84]. In some of the approaches for prosody modification, the speech signal is represented in a parametric form, as in the Harmonic plus Noise Model (HNM), Speech Transformation and Representation using Adaptive Interpolation of weiGHTed spectrum (STRAIGHT) and sinusoidal modeling [72, 85–88, 95]. Pitch modification based on Discrete Cosine Transform (DCT) incorporates the required pitch modification by modifying the LP residual [90]. Some approaches use phase vocoders for time-scale modification [77].

2.3.1 Overlap and add approach

In the overlap and add approach for time-scale modification, the speech signal is split into short (about 2 to 3 pitch periods long) segments using overlapping analysis windows [83]. Each segment is multiplied with a Hann window. For synthesis, the windowed segments are overlapped and added. Based on the desired time-scale modification, some of the windowed segments are either replicated or omitted. In these cases, the information about the pitch markers is not used for splitting the speech signal into short segments. Hence the periodicity due to pitch was not preserved well after the time-scale modification. The synchronous overlap and add (SOLA) approach allows flexible positioning of the windowed segments by searching for the placement of the analysis window in such a way that the overlapped regions have maximum correlation [84].

2.3.2 Pitch synchronous overlap and add approach

While the OLA and SOLA approaches are limited to time-scale modification, the PSOLA approach can be applied to both time and pitch-scale modification [73, 75]. There are several versions of the PSOLA algorithm [75, 96]. The time-domain version, called TD-PSOLA, is most commonly used due to its computational efficiency [97]. The basic method consists of deriving pitch synchronous analysis segments, using pitch markers [73, 98]. The pitch markers can be determined by using a pitch extraction algorithm [97]. Analysis windows are typically of length 2 or 4 pitch periods, and are centered around the pitch marker.

Manipulation of the pitch is achieved by changing the time intervals between the pitch markers. Fig. 2.5 depicts the modification of pitch using TD-PSOLA approach. The modification of duration is achieved by either repeating or omitting the speech segments. The TD-PSOLA suffers from spectral and phase distortions due to direct manipulation of the speech signal. Other variations of PSOLA, namely, Frequency Domain PSOLA (FD-PSOLA) and Linear Prediction PSOLA (LP-PSOLA), are theoretically more suitable for pitch-scale modification, because they provide independent control over the spectral envelope for synthesis. The FD-PSOLA is used only for pitch-scale modification. The LP-PSOLA is used for both pitch and duration modification using the principle of residual excited vocoders [75].

2.3.3 Harmonic plus noise model based approach

In the Harmonic plus Noise Model (HNM), the speech signals are represented as a time-varying harmonic component plus a modulated noise component [85, 86]. The decomposition of speech into these two components allows for more flexible and natural-sounding modification of the prosody parameters of the speech signal.

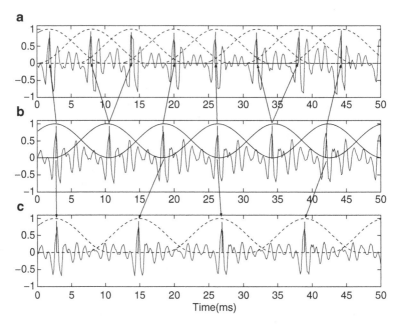

Fig. 2.5 Pitch period modification using TD-PSOLA method. (a) Modified speech signal using TD-PSOLA method for pitch period modification factor $\alpha = 0.75$. (b) A segment of voiced speech (original). (c) Modified speech signal using TD-PSOLA method for pitch period modification factor $\alpha = 1.5$.

But estimation of relevant parameters such as the fundamental frequency (F_0), the maximum voiced frequency, and synthesis time instants involve complex computations. Moreover, the method requires some postprocessing to reduce the interframe incoherence problem (phase mismatch between frames from different acoustic units) [86].

2.3.4 STRAIGHT approach

In the STRAIGHT approach the speech signals are manipulated based on pitch-adaptive spectral smoothing and instantaneous-frequency-based F_0 extraction [87, 88]. In this approach the speech signal is represented using a sequence of F_0 values and the pitch synchronous spectral envelope. The speech parameters are adjusted according to the desired modification either for speech rate modification or F_0 modification. This approach offers greater flexibility than the HNM approach for parameter manipulations without introducing the artificial timbre specific to synthetic speech signals [88]. But this approach requires estimation of instantaneous F_0 and smoothing of the F_0 trajectory.

2.3.5 Sinusoidal model based approach

A different approach for prosody modification is adopted in sinusoidal modeling [72]. In the sinusoidal modeling the speech signal is characterized by amplitudes, frequencies and phases of the component sine waves as shown in Fig. 2.6. These pa-

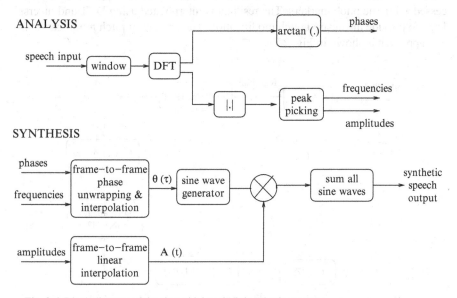

Fig. 2.6 Block diagram of the sinusoidal analysis/synthesis system.

rameters are estimated from the short time Fourier transform of speech. For a given frequency track, a cubic function is used to interpolate the phase as per the desired prosody parameters. This modified phase function is applied to a sine wave generator. The outputs of each of the sine wave generators is amplitude modulated, and is added to similar outputs of the sine wave generators for the other frequency tracks to produce the desired prosody modification. Problems arise when changing the pitch by a large scale factor. In particular, hoarseness was perceived in the reconstruction, when F_0 was increased [99].

2.3.6 DCT based approach

The Discrete Cosine Transform (DCT) based approach performs the pitch modification in the residual domain [90]. For a given speech signal, the Linear Prediction (LP) residual is extracted using the LP analysis. Pitch markers are computed using the autocorrelation function of the residual. The residual in each pitch period is accessed using the pitch markers. The residual is interpolated using DCT and inverse DCT to provide the desired pitch modification. The process of pitch modification in this approach is shown in Fig. 2.7.

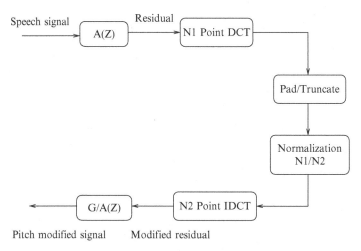

Fig. 2.7 Block diagram for pitch modification using DCT in the source (LP residual) domain.

2.3.7 Phase vocoder

The digital phase vocoder relates the amplitudes and frequencies of the outputs of a digital filterbank to the properties of excitation and vocal tract. The refined phase vocoder proposed by Portnoff takes advantage of the computational efficiency of the fast Fourier transform for implementation. However phase distortions due to pitch modification occur in the synthetic speech signal with objectionable reverberant quality [77].

2.4 Motivation for the Research Work Presented in this Book

The objective of this work is to address certain issues in acquisition and incorporation of duration and intonation knowledge for speech systems in Indian languages.

For acquisition of duration knowledge, rule-based and statistical models have mainly been used in the literature. Rule-based models fail to represent the interactions among the linguistic features, and statistical models suffer with data imbalance problem. Since, nonlinear models are known to capture the implicit relation between the input and output, these models are explored for capturing the implicit duration knowledge present in the speech signal. Among the several existing nonlinear models, Neural Networks (NN) and Support Vector Machines (SVM) are chosen for this task. Neural network models are known for their ability to capture the functional relation between input-output pattern pairs [100, 101]. SVM can provide good generalization on function approximation problems, using the concept of structural risk minimization [100, 102, 103].

For modeling the intonation patterns, most of the researchers have explored phonological models, phonetic models and their variants. All these models generate a sequence of abstract intonational events by analyzing the linguistic content of the text with respect to syntactic and morphological characteristics. These models require a extensive textual analysis and expert linguistic knowledge. For Indian languages, no systematic studies on linguistic analysis have been reported for intonation analysis so far. The existing popular phonetic or phonological models are not suitable for intonation analysis [13, 70, 71]. Nonlinear models such as neural networks and support vector machines are particularly useful in the case of less researched languages (with respect to intonation analysis) like Indian languages, for which most relevant features that affect the intonation patterns and the way they are inter-related have not been studied in detail. Therefore in this work we used NNs and SVMs for modeling the intonation patterns.

Most of the approaches for prosody modification suffer from spectral and phase distortions. This is mainly due to manipulation of the speech signal directly. Some approaches allow the modification of prosody parameters in a limited range, beyond which they introduce distortion. This shows lack of flexibility in incorporating the prosody parameters. Some approaches provides prosody modification at the cost of computational complexity. In this work, a method for prosody modification is proposed, which modifies the excitation source information (linear prediction residual) using the instants of significant excitation of the vocal tract system during the production of speech. This method provides minimum distortion because it operates only in the residual (linear prediction (LP) residual) domain. Since the samples in the LP residual signal are less correlated compared to the samples in the speech signal, modifying the LP residual introduce less distortion in the modified signal.

2.5 Summary

This chapter provides a brief review of approaches for acquisition and incorporation of prosody knowledge for speech systems. Duration models from rule-based and statistical approaches were reviewed. General intonation models that belong to phonological and acoustic-phonetic models were discussed. Some of the existing prosody modification approaches were discussed. Summary of the review of acquisition and incorporation of prosody knowledge for speech systems is given in Tables 2.1 and 2.2, respectively. The next chapter presents the analysis of durations of sound units for the Indian language Telugu.

Table 2.1 Summary of the review of acquisition of prosody knowledge for speech systems.

Acquisition of prosody knowledge: Capturing duration and intonation knowledge using modeling techniques. **Duration knowledge:** Durations of sound units usually modeled using rule-based and statistical methods.

- Rule-based models: Rules derived using expert knowledge.

    - Klatt duration model: Based on information present in phonetic literature.
    - IITM duration model: Based on the analysis of positional and contextual factors.

- Statistical models: Models derived from large data. These models can be based on parametric or nonparametric regression models.

    - Parametric regression model: The structure of processing the input parameters is determined apriori.
        · Sums-of-Products model: Based on prior phonetic and phonological information as well as information collected by analyzing data.
    - Nonparametric regression model: The structure of processing the input parameters is not known.
        · CART model: A binary branching tree is derived by satisfying the constraints.
        · Neural network model: Used to capture the implicit relations present in data.

Intonation knowledge: Intonation patterns can be modeled either by using some phonological rules or by using data-based models.

- Phonological model (Tone sequence model): Intonation patterns are represented by sequence of tones.
- Acoustic-phonetic model (Fujisaki model): Interpret F_0 contour as the result of the superposition of several components of different temporal scopes.
- Perception model (IPO model): Approximating the F_0 contour with perceptually relevant movements.
- Acoustic stylization model (Tilt model): F_0 contour is represented with a sequence of events. Each event is described by a set of continuous parameters.
- Neural network model: F_0 values are predicted using FFNN and RNN models by presenting the contextual, phonological and linguistic features as input to the model.
- IITM intonation model: Rules are derived by performing the analysis at different levels (macro prosody: pitch accents, and micro prosody: prosody variations at segmental level).

Table 2.2 Summary of the review of incorporation of prosody knowledge for speech systems.

Incorporation of prosody knowledge: The acquired prosody knowledge can be incorporated in speech systems by modifying the prosody parameters of speech.

- Overlap and add approach: Windowed speech segments are overlapped and added.
- Pitch synchronous overlap and add approach: Pitch synchronous windowed speech segments are overlapped and added.
- Harmonic plus noise model based approach: Modification of prosody parameters is performed separately on harmonic and noise components of a speech signal.
- STRAIGHT approach: Modification is based on pitch-adaptive spectral smoothing and instantaneous-frequency-based F_0 extraction.
- Sinusoidal model based approach: Speech signal is characterized by a sequence of sine wave components. Prosody modification is performed by manipulating these sine wave components.
- DCT based approach: Modifying the LP residual using DCT and inverse DCT.
- Phase vocoder: Modification of filterbank coefficients, which represents the excitation and vocal tract characteristics of speech.

Chapter 3
ANALYSIS OF DURATIONS OF SOUND UNITS

Abstract This chapter presents the detailed analysis of durations of sound units. Durations of the syllables are analyzed with respect to positional and contextual factors. For detailed duration analysis, syllables are categorized into groups based on size of the word and position of the word in the utterance, and the analysis is performed separately on each category. From the duration analysis presented in this chapter, it is observed that durations of sound units depend on several factors at various levels, and it is very difficult to derive the precise rules for accurate estimation of durations. Therefore, there is a need to explore nonlinear models to capture the duration patterns of sound units from features mentioned in this chapter.

3.1 Introduction

Acoustic analysis and synthesis experiments have shown that duration and intonation patterns are the two most important prosodic features responsible for the quality of synthesized speech [104]. A good prosody model should capture the durational and intonational properties of natural speech. In this chapter we present the analysis of durations of sound units performed on the broadcast news data for the Indian language Telugu. The results of this analysis will help us in modeling the duration and intonation patterns for the sequence of syllables.

This chapter is organized as follows: Discussion on factors that affect the duration of a syllable is presented in Section 3.2. The database used for duration analysis is described in Section 3.3. Computation of average durations and their deviations from the base durations for initial and final syllables is discussed in Section 3.4. Section 3.5 describes the analysis of durations of syllables using positional and contextual factors. Detailed analysis is performed in Section 3.6 by categorizing the syllables based on the size of the word and position of the word in the utterance.

K.S. Rao, *Predicting Prosody from Text for Text-to-Speech Synthesis*, SpringerBriefs in Electrical and Computer Engineering, DOI 10.1007/978-1-4614-1338-7_3,
© Springer Science+Business Media New York 2012

3.2 Factors Affecting the Syllable Duration

The factors affecting the durations of the basic sound units in continuous speech can be broadly categorized into phonological, positional and contextual [105–113]. The vowel is considered as the nucleus of a syllable, and consonants may be present on either side of the vowel. The duration of a syllable is influenced by the position of the vowel, the category of the vowel and the type of the consonants associated with the vowel [106–108]. The positions that affect the durations of the syllables are: Word initial position, word final position, phrase boundary and sentence ending position [114]. The contextual factors include the preceding and the following syllables [109, 110]. The manner and place of articulation of the syllables in the preceding and the following positions also affect the duration of the present syllable [111, 112]. In addition, the gender of the speaker, psychological state of the speaker (happy, anger, fear etc.,), age, relative novelty in the words and words with relatively large number of syllables, also affect the duration [105].

3.3 Speech Database

The database for this study consists of 20 broadcast news bulletins in Telugu. These news bulletins are read by male and female speakers. The total duration of speech in the database is 4.5 hours. The speech signal was sampled at 16kHz and represented as 16 bit numbers. The speech is segmented into short utterances of duration of around 2 to 3 seconds. The speech utterances are manually transcribed into text using common transliteration code (ITRANS) for Indian languages [115]. The speech utterances are segmented and labeled manually into syllable-like units. Each bulletin is organized in the form of syllables, words and orthographic text representations of the utterances. Each syllable and word file contains the text transcriptions and timing information in number of samples. The database consists of 6,484 utterances with 25,463 words and 84,349 syllables [116].

In this work, syllable is chosen as the basic unit for the analysis. The syllable is a natural and convenient unit for speech in Indian languages. In Indian scripts characters generally correspond to syllables. A character in an Indian language script is typically in one of the following forms: V, CV, CCV, CCVC and CVCC, where C is a consonant and V is a vowel. Syllable boundaries are more precisely identified than phonemic segment in both the speech waveform and in the spectrographic display [117].

3.4 Computation of Durations

In order to analyze the effects of positional and contextual factors, syllables need to be categorized into groups based on position and context. Syllables at word initial position, middle position and final position are grouped as initial syllables, middle syllables and final syllables, respectively. Syllables next to initial syllables are grouped as following syllables, while syllables before the final syllables are grouped as preceding syllables. Words with only one syllable are treated as monosyllabic words, and the syllables are known as monosyllables. In Telugu language the occurrence of monosyllables is very less.

To analyze the variations in durations of syllables due to positional and contextual factors, reference duration of the syllable is needed [79]. The reference duration, also known as base duration, the effect of any of the factors should be minimum. For this analysis only the middle syllables are considered as neutral syllables, where the effects of positional and contextual factors are minimum. The base duration of a syllable is obtained by averaging the durations of all the middle syllables of that category. In this analysis, some of the initial/final syllables have no reference duration, because they are not available in the word middle position. This subset of syllables is not considered for analysis.

For analyzing the effect of positional factors, initial and final position syllables are considered. This analysis consists of computation of mean duration of initial/final syllables and their deviation from their base durations. The deviations are expressed in percentage. For analyzing the gross behavior of positional factors, a set of syllables is considered, whose frequency of occurrence is greater than a particular threshold (threshold = 10) across all bulletins. This set consists of about 60 to 70 syllables, and is denoted as the set of common syllables. Most of these syllables are terminated with vowels and a few are terminated with consonants. Table 3.1 shows the percentage deviations of durations of the initial syllables terminating with vowels. Table 3.2 shows the percentage deviations of durations of the final syllables terminating with vowels. In these tables, the leftmost column indicates the consonant part of the syllable (CV or CCV) and the top row indicates the vowel part of the syllable. The other entries in the tables represent the percentage deviations of durations of the syllables. The blank entries in the tables correspond to syllables whose frequency of occurrence is less than a threshold (threshold = 10) across all bulletins.

Contextual factors deal with the effect of the preceding and the following unit on the current unit. In this analysis, middle syllables are assumed as neutral syllables. Therefore, initial and final syllables are analyzed for contextual effects. For initial syllable, only the effect of the following unit is analyzed. For final syllable, only the effect of the preceding syllable is analyzed. To perform the analysis, the following and preceding syllables need to be identified. The percentage deviation of duration is computed for all initial and final syllables. For each following syllable, the mean of the percentage deviation of durations of all corresponding initial syllables is computed. Likewise for each preceding syllable, the mean of the percentage deviation of durations of all corresponding final syllables is computed. These average deviations represent the variations in durations of initial and final syllables due to their following and preceding units, respectively. Table 3.3 shows the percentage deviations of durations of the initial syllables due to their following syllables terminating with

Table 3.1 Percentage deviations of durations of the initial syllables terminating with vowels. The entries in the leftmost column and top row indicate the consonant and vowel parts of the syllable.

	a	A	i	I	u	U	e	E	o	O
k	23	14	31		20	3		14	23	27
ch			32				1	3	10	
t	-2	19	2	11			29	0		
p	5	2	18	9	2		14	24	22	31
g	103	45			70					67
j	37	44								
d	51	36	42					24		
b	71	31	81	60						
bh	40	48								
m	51	44	56		34	48	61			
n	51	15	48	37				40		
r	47	61	58		60			40		8
v	40	39	33	62			40	34		
s	17	15								
sh		19								

Table 3.2 Percentage deviations of durations of the final syllables terminating with vowels. The entries in the leftmost column and top row indicate the consonant and vowel parts of the syllable.

	a	A	i	I	u	U	e	E	o	O
k	28		54	55						
ch			90				28			
T	17		54	45						
t	34	25	46		53	33				12
g		18			62					
j		18	54	51						
D	26		84	53						
d			65	49						
n	22		52	44						
l	17	0			58					31
y	37	-7	53							
r	20		43		62					
v	22									
Sh	13		5							

vowels. Table 3.4 shows the percentage deviations of durations of final syllables due to their preceding syllables terminating with vowels.

Table 3.3 Percentage deviations of durations of initial syllables due to their following syllables terminating with vowel. The syllables indicated are following syllables. The entries in the leftmost column and top row indicate the consonant and vowel parts of the following syllable.

	a	A	i	I	u	U	e	E	o	O
k	32			-1						
ch	34	36	9							
T	30		19	18	21					
t	58	27	28	58	17					
p	21	8	49		34			22		
g	30	21	29		32					
j	29				10					
D	16	4	17		23					
d	17		9	8	15					
b			12		3					
bh	11									
m	11	21	20		21	23				22
n	23	15	15		18					
y	9	4	-1		-7					
r	35	35	21		28					
l	25	30	33	27	25			24		13
v	7	16	8	11				21		
s	8	14	10		5			-1		
Sh	13	4								

Table 3.4 Percentage deviations of durations of the final syllables due to their preceding syllables terminating with vowel. The syllables indicated are preceding syllables. The entries in the leftmost column and top row indicate the consonant and vowel parts of the preceding syllable.

	a	A	i	I	u	U	e	E	o	O
k	11	20			21					18
ch	13	17	21					22		
T	28	20	22		29					
t	14	27	26	26	20					
p	22	26	19		28			26		-3
g	32	22	39		41					
j	37				32					
D	31	23	34		39					
d	35	23	29		33			29		
m	31	26	28							
n	28	33	30							
l	41		34		52					47
y	24	27	25		28			34		
r	32	32	36		33					
v	18	32	28		28					
s	6	8	14							
Sh	8		1							

3.5 Analysis of Durations

Durations of the syllables are analyzed using positional and contextual factors. The effect of positional factors is analyzed by observing the durations of initial and final syllables. The effect of contextual factors is analyzed by observing the durations of initial and final syllables with respect to their following and preceding syllables. The following subsections summarize the effects of positional and contextual factors on syllable duration.

3.5.1 Positional factors

From Table 3.1, it is observed that most of the syllables at word initial position have durations more than their base durations. The percentage deviations of durations of all the initial syllables are not uniform. They vary based on manner of articulation, place of articulation and voicing nature associated with the production of the syllable. At a primary level, it is noticed that syllables with voicing nature (consonant within a syllable is of voicing nature) have more deviations in durations compared to their unvoiced counterparts. Again, within the voiced and unvoiced categories, a variation in duration is observed based on the manner and place of articulation and on the nature of vowel present in the syllable.

From the analysis of word final position syllables (Table 3.2), it is observed that most of the syllables terminating with vowels (i.e., CV type) have larger duration compared to their base duration. Bilabial stops, bilabial nasals and fricative group of syllables do not belong to the set of common syllables. From the final syllables of the set of common syllables, a broad grouping can be performed based on the vowel inside the syllable. Table 3.2 shows that syllables with the vowel /a/ have deviations in duration between 20% and 30%, while syllables with vowel /i/ and /u/ have about 40% to 60% deviations.

3.5.2 Contextual factors

The effect of contextual factors on the initial and final syllables is given (in the form of percentage deviations of durations of syllables) in Tables 3.3 and 3.4, respectively. The initial syllable duration is close to its reference duration in the case of syllables with semivowels or fricatives in following position. The duration of initial syllable increases by 10% to 20% of its base duration, when nasalized syllable is the following unit. Trills and liquids increase the duration of their preceding syllables by 25% to 35%. Syllables with unvoiced stops at the following position affect the durations of their preceding syllables (initial syllables) more compared to syllables with voiced stops at the following position.

The final syllable duration increases by 20% to 30%, if the preceding syllable contains unvoiced stop consonant or semivowel. Syllables with voiced stop consonants and trills affect the duration of the following final syllables by 30% to 40%. The final syllable duration is increased by 45%, if liquid category syllables are in the preceding position. Nasals in the preceding position increase the duration of final syllables by approximately 30%. Fricative based syllables increase the duration of the following syllables by 10%. The important contextual effect observed is that syllables with unvoiced stop consonants affect the durations of their preceding syllables more compared to their following syllables, whereas syllables with voiced stop consonants affect the durations of their following syllables more.

3.6 Detailed Duration Analysis

In the analysis (Section 3.5), a wide range of durational variations are observed in both initial and final syllables. This is due to the dependency of syllable duration on size of the word and position of the word in the utterance. Hence, for a detailed duration analysis, the initial/final syllables need to be categorized further based on word size and position, and the analysis needs to be performed separately on different categories. Analysis of durations of the syllables with respect to position of the word and size of the word is performed in the following subsections.

3.6.1 Analysis of durations based on position of the word

To perform this analysis, words are classified into groups based on their position in the utterance. The word positions considered for the analysis are first, middle, and last, denoted by W_f, W_m, and W_l, respectively. From each group of words the following analysis is performed: Initial and final syllables, their adjacent syllables and their associated durations are derived. The average deviations of the durations of the initial and final syllables are computed using positional and contextual factors. The set of syllables present in each category above some threshold of frequency of occurrence is used for analysis. Table 3.5 shows the percentage deviations of durations of initial and final syllables present in the words at different positions in the utterance. Table 3.6 shows the percentage deviations of durations of the initial and final syllables due to their associated context (following and preceding syllables) present in the words at different positions in the utterance.

The following are the inferences drawn from the Table 3.5

- Initial syllables present in W_m have more duration compared to initial syllables in W_f and W_l.
- Initial syllables with unvoiced consonants present in W_f have durations lesser than their base durations.

Table 3.5 Percentage deviations of durations of initial and final syllables present in the words that occur at different positions in the utterance

Initial syllables				Final syllables		
	first	middle	last		first	middle	last
a	7	46	30	Du	38	41	95
ba	47	85	51	TI	33	19	47
da	47	55	43	Tu	44	21	98
ga	94	113	77	chi	80	79	104
ka	-8	43	34	du	43	29	71
na	62	54	52	ga	54	27	115
pa	-25	22	13	ju	37	37	77
reN	32	49	51	ka	28	19	82
ta	-29	24	17	ku	37	34	87
bA	72	59	38	la	14	9	68
dE	45	39	28	lu	49	46	88
ja	64	61	37	na	22	14	48
kAr	-3	11	20	pu	18	11	72
nA	47	42	13	ra	25	8	79
rA	68	62	53	ram	10	1	14
vi	20	35	40	sham	7	1	20
mukh	11	33	57	ta	26	22	76
rASh	17	25	26	tri	42	31	52
				vam	5	-3	16
				ya	35	22	97

- Initial syllables from nasal and trill categories present in W_f have greater durations compared to other word positions.
- The initial syllables terminating with long vowels have more duration in W_f, whereas the initial syllables terminating with consonants have more duration in W_l.
- The final syllables of W_l have larger duration compared to final syllables of W_f and W_m, and among the syllables of W_f and W_m, syllables of W_f have larger durations.
- In comparison with the initial syllables of W_l, the final syllables of W_l have larger deviations in durations.

The following are the inferences drawn from the Table 3.6

- The initial syllables present in W_m and W_l are more lengthened due to their following syllables.
- The final syllables of W_f and W_l are more lengthened compared to the initial syllables of W_f and W_l due to their context.
- The durations of the final syllables of W_l are more lengthened due to their preceding syllables, compared to other word positions.

Table 3.6 Percentage deviations of durations of initial and final syllables due to their adjacent syllables (following and preceding syllables for initial and final syllables, respectively) in the words that occur at different positions in the utterance. Syllables in the first column are following syllables to the initial syllables and syllables in the fifth column are preceding syllables to the final syllables.

Following syllables				Preceding syllables			
	first	middle	last		first	middle	last
Du	35	45	49	Da	30	27	34
TI	-7	14	10	Sha	9	4	24
bhut	0	31	32	Ta	20	13	46
da	6	16	28	bhut	42	43	64
di	-6	16	13	da	21	20	63
ga	40	32	22	dhA	4	2	44
ju	38	52	62	du	19	8	68
la	11	31	22	ga	14	1	43
lu	0	34	35	la	29	27	61
ma	-15	35	33	ma	24	13	67
na	13	25	31	man	41	28	50
ni	6	21	8	na	28	16	49
ra	22	45	34	ni	19	66	72
ru	6	39	36	pa	36	6	41
ta	52	64	59	ra	17	23	67
va	28	27	25	ri	31	28	58
ya	-4	12	14	sa	8	1	34
				ta	12	13	25
				tu	27	24	51
				va	17	6	43
				ya	29	12	63

3.6.2 Analysis of durations based on size of the word

In this analysis, words are categorized into groups based on the number of syllables they contain. In this study words are classified into six groups. They are monosyllabic, bisyllabic, trisyllabic, tetrasyllabic, pentasyllabic and polysyllabic words, containing of one, two, three, four, five and more than five syllables, respectively. Monosyllabic words are not considered for analysis, since they are very few in number. Analysis is performed separately for the other five groups. Table 3.7 shows the percentage deviations of durations of the initial syllables and final syllables present in different word sizes. Table 3.8 shows the percentage deviations of durations of the initial and final syllables due to their adjacent syllables.

The results of the analysis indicates that, the durations of the initial and final syllables in different word sizes are inversely related to word size. That is, the durations of the initial/final syllables in bisyllabic words are more compared to the durations of the initial/final syllables in polysyllabic words. The mean durations of the final

Table 3.7 Percentage deviations of durations of initial and final syllables present in different word sizes.

	Initial syllables						Final syllables				
	bi	tri	tetra	penta	poly		bi	tri	tetra	penta	poly
a	41	37	29	27	34	Du	46	41	61	59	33
chE	13	4	3	3	3	Ti	68	54	29	31	24
ka	52	20	30	30	13	chi	93	83	94	88	82
ku	30	21	21	18	16	da	80	26	17	23	-5
ma	48	52	52	51	46	gu	64	20	-1	10	24
nA	31	8	16	6	31	ka	37	24	21	23	47
na	76	71	28	51	40	la	59	17	8	13	13
ni	61	45	50	38	49	nu	50	43	40	41	37
pa	28	10	0	6	2	ra	52	6	15	-6	18
ra	27	21	23	21	21	ru	45	24	19	15	24
sa	19	17	19	11	19	si	61	34	38	11	-5
tI	22	13	4	19	8	ti	38	29	43	21	25
vi	42	35	32	33	32	ya	69	25	27	30	27
kAr	27	26	-1	-7	-7						

Table 3.8 Percentage deviations of durations of initial and final syllables due to their adjacent syllables (following and preceding syllables for initial and final syllables, respectively) present in different word sizes. Syllables in the first column are following syllables to the initial syllables and syllables in the seventh column are preceding syllables to final syllables.

	Following syllables						Preceding syllables				
	bi	tri	tetra	penta	poly		bi	tri	tetra	penta	poly
Du	52	57	45	44	30	chA	10	24	21	13	36
Ta	25	21	21	13	16	da	-25	30	30	32	49
da	45	15	16	18	3	gA	27	17	20	32	39
di	22	-2	14	-10	7	ku	-22	24	41	23	33
li	51	43	23	12	40	la	28	24	47	41	38
lu	30	26	23	21	22	ma	12	14	40	56	62
mi	38	20	21	16	-11	man	32	39	40	27	43
nu	22	7	19	25	11	nA	23	24	20	20	33
ni	37	35	16	5	15	na	11	15	30	30	54
pu	53	13	13	-8	40	ni	27	0	24	43	58
ra	46	45	25	43	24	rA	9	26	35	33	48
ri	30	18	25	18	9	ra	34	18	43	40	68
ru	51	22	31	39	27	shA	20	13	33	27	27
si	25	8	4	10	-2	vA	26	27	36	43	49
su	12	4	7	8	0	yA	19	36	35	28	41
ta	70	50	47	65	52						
ti	34	36	27	24	19						
va	51	24	23	31	22						
ya	16	5	8	10	10						

syllables from various word sizes are more, compared to the initial syllables of their corresponding categories.

The deviations of durations of the initial and final syllables due to contextual factors are less in magnitude, compared to deviations in durations due to positional factors. The percentage deviations of durations of the initial syllables due to their following units are inversely related to size of the word, whereas for the final syllables, the deviations in durations due to their preceding units are proportional to size of the word.

So far duration analysis is performed on the whole set of initial/final syllables, and on the categorized initial/final syllables based on position/size of the word in the utterance. In the analysis a large variation in duration within each category of syllables is observed. For more detailed analysis, there is a need to classify the syllables further by considering the size of the word and position of the word together. With this classification, words can be categorized into 15 groups ($3 \times 5 = 15$, using position of the word 3 groups, and size of the word 5 groups).

3.7 Observation and Discussion

From the above analysis (Section 3.5 and Section 3.6) it is observed that duration patterns for the sequence of syllables depends on different factors at various levels. It is difficult to derive a finite number of rules which characterizes the behaviour of the duration patterns of the syllables. Even to fit the linear models to characterize the durational behaviour of the syllables is also difficult [118]. Since the linguistic features associated to different factors have complex interactions at different levels, it is difficult to derive a rulebase or linear model to characterize the durational behaviour of the syllables.

To overcome the difficulty in modeling the duration patterns of the syllables, nonlinear models can be explored. Nonlinear models are known for their ability to capture the complex relations between the input and output. The performance of the model depends on the quality and quantity of the training data, structure of the model and training and testing topologies [100]. From the analysis carried out in sections 3.5 and 3.6, we can identify the features that affect the duration of a syllable. Nonlinear models can be developed using these features as input and the corresponding duration as the output to predict the durations of syllables [2, 119, 120]. Similar analogy can be used for modeling the intonation patterns [3, 67].

Even though the list of the factors affecting the duration may be identified, but it is difficult to determine their effect independently. This is because the duration patterns of the sequence of sound units depend on several factors and their complex interactions[121, 122]. Formulation of these interactions in terms of either linear or nonlinear relations is a complex task. For example, in the analysis of positional factors (Tables 3.1 and 3.2) the deviation in the durations include the effect of contextual factors, nature of the syllable, word level and phrase level factors. Similarly in the analysis of contextual factors, the contributions of other factors are also included.

From the duration analysis one can identify the list of features affecting the duration, but it is very difficult to derive the precise rules for estimating the duration. The analysis performed in this chapter may be useful for some speech applications where the precise estimation of durations is not essential. For example, in the case of speech recognition, rule-based duration models can provide a supporting evidence to improve the recognition rate [123]. This is particularly evident in speech recognition in noisy environment [124]. In the case of speaker recognition, speaker-specific duration models will give an additional evidence, which can be further used to enhance the recognition performance [125]. Duration models are also useful in language identification task, since the duration patterns of the sequence of sound units are unique to a particular language [126–128].

In the duration analysis, the numbers shown in the tables indicate the average deviations. These models may not be appropriate for Text-to-Speech (TTS) synthesis application. In TTS synthesis, precise duration models produce speech with high naturalness and intelligibility [10, 129]. The naturalness mainly depends on the accuracy of prosody models. The derived duration models in the paper may not be appropriate for high quality TTS applications, but they can be useful in developing other speech systems such as speech recognition, speaker recognition and language identification. Precise duration models can be derived by using nonlinear models such as neural networks, support vector machines and classification and regression trees. A nonlinear model will give the precise duration by providing (1) all the factors and features responsible for the variation in duration as input, (2) each category of the sound unit has enough examples and (3) the database should contain enough diversity [100].

3.8 Summary

Factors affecting the durations of syllables in continuous speech were identified. They are positional, contextual and phonological factors. Durations were analyzed using positional and contextual factors. In the analysis of positional factors, it was noted that the deviations in durations of the syllables depend on voicing, place and manner of articulation, and nature of vowel present in the syllable. From the analysis of contextual factors, it was mainly observed that syllables with unvoiced stop consonants affect the durations of their preceding syllables, and syllables with voiced consonants affect the durations of their following syllables. In the analysis, a wide range of durational variations was observed. For detailed analysis, categorization of syllables was suggested. In this work syllables were categorized based on size of the word and position of the word in the utterance, and the analysis was performed separately in each group.

Chapter 4
MODELING DURATION

Abstract This chapter presents modeling of durations of syllables. Features representing linguistic and production constraints are proposed for modeling the durations of syllables. Feedforward Neural Network (FFNN) and Support Vector Machines (SVMs) are proposed for predicting the durations of syllables. The prediction accuracy of the proposed models is compared with Classification and Regression Tree (CART) models. The effect of different constraints and their interactions in modeling the durations of syllables is examined. Finally, the proposed duration models have demonstrated for capturing the language-specific and speaker-specific information.

4.1 Introduction

In the previous chapter durations of the syllables are analyzed by computing the average durations and deviations from their reference durations. The effect of positional and contextual factors was analyzed separately by grouping the syllables based on size of the word and the position of the word in the utterance. This chapter discuss the modeling of durations of syllables using neural networks and support vector machines. The implicit knowledge of prosody is usually captured using modeling techniques. In this chapter we focus on modeling the duration knowledge in speech. In speech signal, the duration of each sound unit is dictated by the linguistic context of the unit and the production constraints.

The objective in the present study is to determine whether nonlinear models can capture the implicit knowledge of the syllable duration in a language. One way to infer this is to examine the error for the training data. If the error is reducing for successive training cycles, then one may infer that the network is indeed capture the implicit relations in the input-output pairs. In this chapter the ability of models to capture the duration knowledge for speech is examined in different Indian

K.S. Rao, *Predicting Prosody from Text for Text-to-Speech Synthesis*, SpringerBriefs
in Electrical and Computer Engineering, DOI 10.1007/978-1-4614-1338-7_4,
© Springer Science+Business Media New York 2012

languages. Here three Indian languages (Hindi, Telugu and Tamil) are considered
with syllable as the basic sound unit for analyzing the prediction performance of the
proposed duration models.

The prediction performance of the neural network model depends on the nature
of training data used. Distributions of the durations of syllables (Fig. 4.1) indicate
that majority of the durations are concentrated around mean of the distribution. The
distribution of the syllable durations for the three languages is shown in Fig. 4.1.
This kind of training data forces the model to be biased towards mean of the dis-

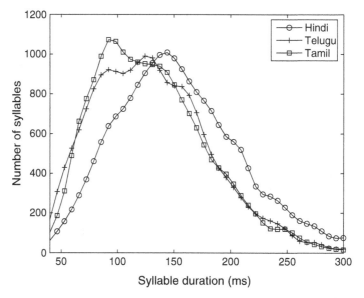

Fig. 4.1 Distributions of the durations of syllables in the languages (Hindi, Telugu and Tamil).

tribution. To avoid this problem, some preprocessing and postprocessing methods
are proposed. Preprocessing methods involve use of multiple models, one for each
limited range of duration. This requires a two-stage duration model, where the first
stage is used to segregate the input into groups, and the second stage is used for
prediction. Postprocessing methods modify the predicted values further using some
durational constraints.

This chapter is organized as follows: Section 4.2 discusses the features used as
input to the neural network and SVM for capturing the knowledge of the syllable
duration. The database used for the proposed duration model is described in Section 4.3. Section 4.4 gives the details of the neural network model, and discusses
its performance in predicting the durations of the syllables. Section 4.5 presents the
architecture of the SVM regression model and its performance in predicting the

durations of the syllables. Section 4.6 discusses postprocessing of the predicted values of the duration using the knowledge of the constraints associated with the positions of the syllables. A two-stage duration model is proposed in Section 4.7 to reduce the error in the predicted durations of the syllables. Section 4.8 demonstrates the ability of the derived duration models in identifying (discriminating) speakers and languages by capturing the speaker-specific and language-specific information, respectively.

4.2 Features for Developing the Duration Model

In this study we use 25 features (which form a feature vector) for representing the linguistic context and production constraints of each syllable. These features represent positional, contextual and phonological information of each syllable. Features representing the positional information are further classified based on the position of a word in the phrase and the position of the syllable in a word and phrase.

Syllable position in the phrase: Phrase is delimited by orthographic punctuation. The syllable position in a phrase is characterized by three features. The first feature represents the distance of the syllable from the starting position of the phrase. It is measured in number of syllables i.e., the number of syllables ahead of the present syllable in the phrase. The second feature indicates the distance of the syllable from the terminating position of the phrase. The third feature represents the total number of syllables in the phrase.

Syllable position in the word: In Indian languages words are identified by spacing between them. The syllable position in a word is characterized by three features similar to the phrase. The first two features are the positions of the syllable with respect to the word boundaries. The third feature is the number of syllables in a word.

Position of a word: The duration of a syllable may depend on the position of the word in an utterance. Therefore the word position is used for developing the duration model. The word position in an utterance is represented by three features. They are: the positions of the word with respect to the phrase boundaries, and the number of words in the phrase.

Syllable identity: A syllable is a combination of segments of consonants (C) and vowels (V). In this study, syllables with more than four segments (Cs or Vs) are ignored. Each segment of a syllable is encoded separately, so that each syllable identity is represented by four features.

Context of a syllable: Syllable duration may be influenced by its adjacent syllables. Hence for modeling the duration of a syllable, the context information is represented by the previous and following syllables. Each of these syllables is represented by a four dimensional feature vector, representing the identity of the syllable.

Syllable nucleus: Another important feature is the vowel position in a syllable, and the number of segments before and after the vowel in a syllable. This feature is represented with a three dimensional feature vector specifying the consonant-vowel structure present in the syllable.

Gender identity: The database contains speech from both male and female speakers. This gender information is represented by a single feature.

The list of features and the number of input nodes in a neural network needed to represent the features are given in Table 4.1. These features are coded and normalized before presented to the neural network model. The details of coding the features are given in Appendix A.

Table 4.1 List of factors affecting the syllable duration, features representing the factors and the number of nodes needed for neural network to represent the features.

Factors	Features	# Nodes
Syllable position in the phrase	Position of syllable from beginning of the phrase Position of syllable from end of the phrase Number of syllables in the phrase	3
Syllable position in the word	Position of syllable from beginning of the word Position of syllable from end of the word Number of syllables in the word	3
Word position in the phrase	Position of word from beginning of the phrase Position of word from end of the phrase Number of words in the phrase	3
Syllable identity	Segments of the syllable (consonants and vowels)	4
Context of the syllable	Previous syllable	4
	Following syllable	4
Syllable nucleus	Position of the nucleus Number of segments before the nucleus Number of segments after the nucleus	3
Gender identity	Gender of the speaker	1

4.3 Database for Duration Modeling

The database for this study consists of 19 Hindi, 20 Telugu and 33 Tamil broadcast news bulletins. In each language these news bulletins are read by male and female speakers. The total durations of speech in Hindi, Telugu and Tamil are 3.5 hours, 4.5 hours and 5 hours, respectively. The speech signal was sampled at 16kHz and represented as 16 bit numbers. The speech utterances are manually transcribed into text using common transliteration code (ITRANS) for Indian languages [115]. The speech utterances are segmented and labeled manually into syllable-like units. Each bulletin is organized in the form of syllables, words and orthographic text representations of the utterances. Each syllable and word file contains the text transcriptions and timing information in number of samples. The syllable durations vary from 30 ms to 450 ms. The description of the data with respect to duration for the three languages is given in Table 4.2 [116].

Table 4.2 Details of the broadcast news data for the languages Hindi, Telugu and Tamil.

| | # Speakers | | # Utter- | # Words | # Sylla- | Syllable duration (ms) | |
Language	Male	Female	ances		bles	Mean	Std. dev.
Hindi	06	13	4,191	26,090	50,237	157.10	57.37
Telugu	11	09	6,484	25,463	84,349	133.65	54.69
Tamil	10	23	7,359	30,688	1,00,707	132.10	48.84

4.4 Duration Modeling using Feedforward Neural Networks

A four layer Feedforward Neural Network (FFNN) is used for modeling the durations of syllables. The general structure of the FFNN is shown in Fig. 4.2. The first

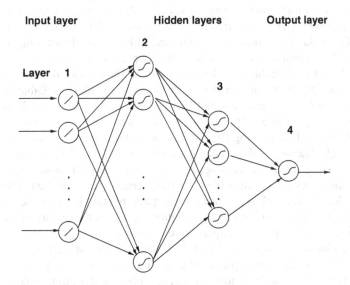

Fig. 4.2 Four layer feedforward neural Network.

layer is the input layer with linear units. The second and third layers are hidden layers. The second layer (first hidden layer) of the network has more units than the input layer, and it can be interpreted as capturing some local features in the input space. The third layer (second hidden layer) has fewer units than the first layer, and can be interpreted as capturing some global features [100, 101]. The fourth layer is the output layer having one unit representing the duration of a syllable. The activation function for the units at the input layer is linear, and for the units at the hidden layers, it is nonlinear. Generalization by the network is influenced by three factors: The size of the training set, the architecture of the neural network, and the complexity of the problem. We have no control over the first and last factors. Several network structures are explored in this study. The (empirically arrived) final structure of the network is $25L\ 50N\ 12N\ 1N$, where L denotes a linear unit, and N denotes a nonlinear unit. The integer value indicates the number of units in that layer. The nonlinear units use $tanh(s)$ as the activation function, where s is the activation value of that unit. For studying the effect of positional and contextual factors on syllable duration, the network structures are given by $14L\ 28N\ 7N\ 1N$ and $13L\ 26N\ 7N\ 1N$, respectively. The inputs to these networks represent the positional and contextual factors. All the input and output features are normalized to the range [-1, +1] before presenting to the neural network. The backpropagation learning algorithm is used for adjusting the weights of the network to minimize the mean squared error at the output for each syllable duration [101].

A separate model is developed for each of the three languages. For Hindi 35000 syllables are used for training the network, and 11222 syllables are used for testing. For Telugu 64000 syllables are used for training, and 17630 are used for testing. For Tamil 75000 syllables are used for training, and 21493 are used for testing. For each syllable a 25 dimension input vector is formed, representing the positional, contextual and phonological features, as described earlier. The duration of each syllable is obtained from the timing information available in the database. Syllable durations seem to follow a logarithmic distribution, and hence the logarithm of the duration is used as the target value [32]. The number of epochs needed for training depends on the behaviour of the training error. It was found that 500 epochs are adequate for this study. The learning ability of the network from training data can be observed from the training error. The training errors for neural network models for each of the three languages are shown in Fig. 4.3. The decreasing trend in the training error indicates that the network is capturing the implicit relation between the input and output. The training error for Hindi is slightly higher, as the quality of the Hindi speech data is slightly poorer compared to the other two.

The duration model is evaluated with the syllables in the test set. For each syllable in the test set, the duration is predicted using the FFNN by presenting the feature vector of each syllable as input to the neural network. The deviation of the predicted duration from the actual duration is obtained. The prediction performance of the models for the three languages is shown in Fig. 4.4. Each plot represents the average predicted duration vs the average duration of a syllable.

For studying the effect of positional and contextual factors on the syllable duration, features associated with the syllable position and syllable context are used separately. Features representing the positional factors are: (a) Syllable position in the phrase (3 dimensional feature), (b) syllable position in the word (3 dimensional feature), (c) word position in the phrase (3 dimensional feature), (d) syllable identity (4 dimensional feature) and (e) identity of gender. Features representing the contextual factors are the identities of the present syllable, its previous and following syllables and identity of gender. The percentage of syllables predicted within different deviations from their actual durations is given in Table 4.3. The first col-

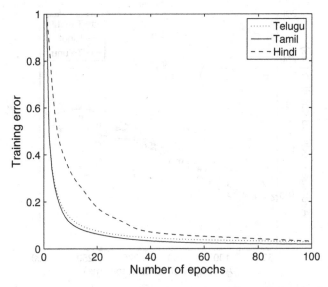

Fig. 4.3 Training errors for the neural network models used for predicting the durations of the syllables in three Indian languages (Hindi, Telugu and Tamil).

Table 4.3 Percentage of syllables predicted within different deviations for different input features for the languages Hindi, Telugu and Tamil.

Language # Syllables	Features	% Predicted syllables within deviation		
		10%	25%	50%
Hindi (11222)	All	29	68	84
	Positional	26	63	81
	Contextual	27	65	82
Telugu (17630)	All	29	66	86
	Positional	25	60	82
	Contextual	26	62	83
Tamil (21493)	All	34	75	96
	Positional	29	70	91
	Contextual	31	72	93

umn indicates the number of syllables specific to the particular language used in testing. The second column shows the features used as input to the neural network. The other columns indicate the percentage of syllables having predicted duration within the specified deviation with respect to their actual durations. Compared to the positional and contextual factors separately, features using all the factors seems to yield a better duration model.

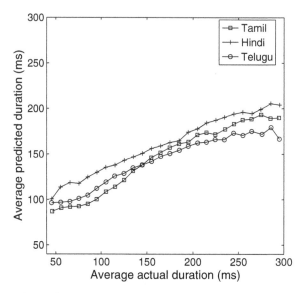

Fig. 4.4 Prediction performance of the neural network models for the languages Hindi, Telugu and Tamil.

In order to evaluate the prediction accuracy, the average prediction error (μ), the standard deviation (σ), and the correlation coefficient ($\gamma_{X,Y}$) are computed using actual and predicted duration values. These results are given in Table 4.4. The defi-

Table 4.4 Performance of neural network models using objective measures (μ, σ and γ).

Language	Avg. prediction error in ms (μ)	Std. dev. in ms (σ)	Corr. coeff. (γ)
Hindi	32	26	0.75
Telugu	29	23	0.78
Tamil	26	22	0.82

nitions of average prediction error (μ), standard deviation (σ), and linear correlation coefficient ($\gamma_{X,Y}$) are given below:

$$\mu = \frac{\sum_i |x_i - y_i|}{N}$$

$$\sigma = \sqrt{\frac{\Sigma_i d_i^2}{N}}, \quad d_i = e_i - \mu, \quad e_i = |x_i - y_i|$$

where x_i, y_i are the actual and predicted durations, respectively, and e_i is the error between the actual and predicted durations. The deviation in error is d_i, and N is the number of observed syllable durations. The correlation coefficient is given by

$$\gamma_{X,Y} = \frac{V_{X,Y}}{\sigma_X \cdot \sigma_Y}, \quad \text{where} \quad V_{X,Y} = \frac{\Sigma_i |(x_i - \bar{x})| \cdot |(y_i - \bar{y})|}{N}$$

The quantities σ_X, σ_Y are the standard deviations of the actual and predicted durations respectively, and $V_{X,Y}$ is the correlation between the actual and predicted durations.

In speech signal, the duration and intonation patterns of the sequence of sound units are interrelated at some higher level through emphasis (stress) and prominence of the words and phrases. But it is difficult to represent the feature vector to capture these dependencies. In this study, the durations of the adjacent syllables are considered as duration constraints, and the intonation patterns (average pitch values) of the present syllable and its adjacent syllables are considered as intonation constraints, for estimating the duration of the syllable. For studying the influence of duration and intonation constraints in predicting the durations of the syllables, separate models are developed with respect to different constraints. They are: (a) model with linguistic features, (b) model with linguistic features and duration constraints, (c) model with linguistic features and intonation constraints and (d) model with linguistic features, duration and intonation constraints. The performance of these models is given in Table 4.5. The results indicate that the prediction performance has improved by imposing the constraints. Better performance is observed when all the constraints are applied together. Improvement in the prediction accuracy is not significant, since the models are trained with less amount of data.

4.5 Duration Modeling using Support Vector Machines

Support Vector Machines (SVMs) perform function approximation using the concept of statistical learning theory [102]. In the training phase, the SVM model finds a global minimum, and tries to maintain the training error close to zero. SVM can provide good generalization on function approximation problems, by implementing the concept of structural risk minimization [100, 130].

Support vector machine predicts the durations of syllables using regression (function approximation). The general SVM architecture for regression is shown in Fig. 4.5. In the training data $\{(x_1, y_1), (x_2, y_2), ... (x_l, y_l)\}$, x_i denotes the input pattern, and y_i is the desired response (target) for the corresponding input pattern. In regression, the goal is to find a function $f(x)$ that gives an output within a prespecified deviation from the desired output, for all the training data. The SVM performs

Table 4.5 Accuracy of prediction by the FFNN models for different languages using different constraints.

Features	Language # Syllables	% Predicted syllables within dev.			Objective measures		
		10%	25%	50%	μ (ms)	σ (ms)	γ
Linguistic	Hindi(1084)	25	57	85	34	27	0.71
	Telugu(1107)	24	56	83	35	28	0.74
	Tamil(949)	26	60	88	30	25	0.73
Linguistic and duration	Hindi(1084)	25	59	86	34	27	0.72
	Telugu(1107)	24	58	85	34	27	0.75
	Tamil(949)	27	62	88	30	25	0.74
Linguistic and intonation	Hindi(1084)	27	60	88	33	27	0.73
	Telugu(1107)	25	58	86	34	27	0.76
	Tamil(949)	28	64	90	29	25	0.75
Linguistic, duration and intonation	Hindi(1084)	29	63	90	32	26	0.73
	Telugu(1107)	27	60	88	33	26	0.77
	Tamil(949)	30	65	92	29	24	0.77

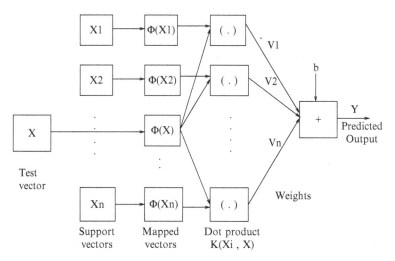

Fig. 4.5 SVM model for regression.

nonlinear transformation for mapping the data onto a high-dimensional feature space. For regression a linear hyperplane is constructed in the feature space such that it is close to majority of the data points [100, 103]. The optimum hyperplane is represented by a subset of training data vectors, which are known as support vectors. The duration of a syllable is predicted using support vectors. In the training data, vectors other than the support vectors will not contribute much for estimating the function. The number of support vectors depends on the precision required for approximating the function from the training data [103, 131]. Here Gaussian kernel Φ is used for nonlinear transformation of data from input space to high-dimensional feature space. For predicting the duration of a syllable, dot products are computed with the images of the support vectors under the mapping Φ. This corresponds to evaluating the inner product kernel functions $K(\mathbf{x_i}, \mathbf{x})$, where $\mathbf{x_i}$, $1 \leq i \leq n$, represent support vectors and $\mathbf{x}$ represents the test vector. Finally the dot products (inner product kernel functions $K(\mathbf{x_i}, \mathbf{x}) = \Phi(\mathbf{x_i}).\Phi(\mathbf{x})$) are added up using weights V_i. This plus the constant bias b yields the final prediction output.

The extracted input vectors representing positional, contextual and phonological features, and the corresponding syllable durations are presented to an SVM model. For the given training data the SVM model is optimized by varying the precision and standard deviation of the Gaussian kernel Φ by trial and error. The performance of the SVM models is given in Table 4.6. Also, the prediction performance of the SVM

Table 4.6 Accuracy of prediction by the SVM regression models for different languages.

Language	# Support vectors	% Predicted syllables within deviation			Objective measures		
		10%	25%	50%	μ (ms)	σ (ms)	γ
Hindi	10249	30	66	84	32.45	26.71	0.74
Telugu	15946	31	68	88	28.21	24.45	0.78
Tamil	20122	33	73	95	26.17	22.39	0.81

regression models for the three languages is shown in Fig. 4.6. Each plot represents the average predicted duration vs the average duration of a syllable. The effect of intonation and duration constraints for predicting the durations of the syllables is analyzed by building separate SVM models. The performance seems to be similar to that of the FFNN models using constraints.

The performance of the FFNN and SVM models can be observed in the Figs. 4.4 and 4.6. The performance characteristics in both the plots seem to be similar. The performance measures (objective measures as well as percentage of syllables predicted within specific deviation) also indicate that both the nonlinear models (FFNN and SVM) perform similarly. The plots indicate that the prediction accuracy is not uniform for the entire range of duration. For improving the accuracy of prediction, postprocessing and preprocessing methods are proposed, and they are discussed in the following sections.

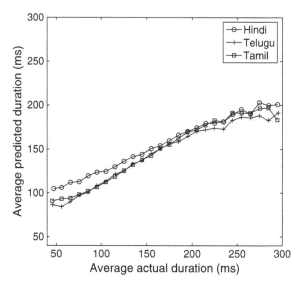

Fig. 4.6 Prediction performance of the SVM regression models for the languages Hindi, Telugu and Tamil.

4.6 Postprocessing of the Predicted Duration Values

The accuracy of prediction (Figs. 4.4 and 4.6) of the duration models is not uniform for the entire range of duration. Better prediction is observed around the mean of the distribution of the original training data. Syllables with long durations tend to be underestimated, and short durations tend to be overestimated. For improving the accuracy of prediction a method is proposed, which modifies the predicted values by imposing piecewise linear transformation on the output predicted durations [132–134]. The piecewise linear transformation is defined by:

$$F(x) = \begin{cases} \alpha x + a(1-\alpha), & 40 \leq x < a \\ x, & a \leq x \leq b \\ \beta x + b(1-\beta), & b \leq x < 300 \end{cases}$$

Here $F(x)$ denotes the transformed predicted value, x is the predicted value obtained from the models, and the parameters $a, b, \alpha,$ and β (all nonnegative) help to control the shape of the function. The interval $[a, b]$ defines the identity portion of the transformation, while α and β control the amount of compression/expansion in the intervals $[40, a]$ and $[b, 300]$, respectively. The bounds 40 and 300 represent the minimum and maximum values of duration in milliseconds. The values $\alpha, \beta < 1$

correspond to compression, and the values $\alpha, \beta > 1$ correspond to expansion. Fig. 4.7 shows the shape of the transformation function for two sets of values:

(1) $a = 100, b = 200, \alpha = 0.3, \beta = 0.5$ and
(2) $a = 100, b = 200, \alpha = 2.0, \beta = 1.8$.

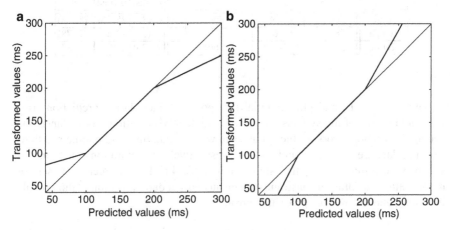

Fig. 4.7 Transformation with (a) compression at both ends ($a = 100, b = 200, \alpha = 0.3$ and $\beta = 0.5$) and (b) expansion at both ends ($a = 100, b = 200, \alpha = 2.0$ and $\beta = 1.8$).

The prediction performance of the models (Figs. 4.4 and 4.6) require the transformation to provide expansion at both the ends for improving the accuracy of prediction. The limits a and b are derived from the distribution of the durations. The optimum parameters of the transformation for the given predicted values for the language Tamil are found to be $a = \mu - 0.45 * \sigma, b = \mu + 0.60 * \sigma, \alpha = 1.25, \beta = 1.55$, where μ and σ are mean and standard deviation of the distribution of syllable durations, respectively. The duration values are recomputed from the predicted values using the optimum transformation. Likewise the transformation is performed on the predicted syllable durations of Telugu and Hindi. The performance measures derived from the transformed and actual duration values are given in Table 4.7. The numbers within brackets indicate the performance measures derived from the predicted (without transformation) and actual duration values. The prediction performance has slightly improved by using the transformation for the entire data. The performance may be further improved by applying the transformation separately for different categories of syllables based on their duration constraints.

The prediction error can be reduced by using the duration constraints of the syllables at different positions (initial, middle and final) in the word and phrase. The prediction error for the syllables in word middle (WM) position is less compared to that of the syllables in word initial (WI) and word final (WF) positions. Syllables

Table 4.7 Performance measures after applying the general piecewise linear transformation. The numbers within brackets indicate the performance measures derived from predicted (without transformation) and actual durations.

Language	% Transformed syllables within deviation			Objective measures		
	10%	25%	50%	μ (ms)	σ (ms)	γ
Hindi	30(29)	69(68)	86(84)	31(32)	25(26)	.76(.75)
Telugu	28(29)	68(66)	87(86)	28(29)	22(23)	.77(.78)
Tamil	34(34)	76(75)	96(96)	26(26)	21(22)	.82(.82)

at phrase final (PF) and phrase initial (PI) positions have larger prediction error compared to that of the syllables in phrase middle (PM) position. For instance, the average prediction errors for the syllables in word initial, final and middle positions in Tamil data are 29.7, 27.8 and 19.8 ms, respectively. The duration constraints of syllables for different languages are given in Table 4.8. The parameters chosen for representing the duration constraints are mean and standard deviation of the syllable durations, and the average prediction error.

Table 4.8 Duration constraints of the syllables for the languages Hindi, Telugu and Tamil.

Language	Parameters	Syllable position in the word			Syllable position in the phrase		
		initial	middle	final	initial	middle	final
Hindi	Mean (μ)	161.9	131.8	172.1	169.2	148.6	177.7
	Std. dev. (σ)	54.8	38.6	59.1	51.2	44.7	53.3
	Avg. error	33.9	23.8	38.8	36.6	28.2	42.8
Telugu	Mean (μ)	153.2	113.3	151.6	161.2	127.3	164.7
	Std. dev. (σ)	51.7	37.1	49.8	47.2	41.5	45.3
	Avg. error	32.1	20.7	30.3	34.6	28.7	37.8
Tamil	Mean (μ)	150	112.5	145.7	158.2	126.4	160.7
	Std. dev. (σ)	48.3	36.4	48	45.2	41.2	47.3
	Avg. error	29.7	19.8	27.8	30.6	26.3	31.8

Using the knowledge of the duration constraints mentioned in the Table 4.8, the optimal parameters (a, b, α, β) are derived for each category of syllables (WI, WM, WF, PI, PM and PF) in the languages Hindi, Telugu and Tamil. Since the prediction error is more for long duration syllables, the associated expansion factor β will be greater than α. The identity portion limits $[a, b]$ are derived from the parameters of the syllable duration distribution. The optimal parameters for different categories of syllables in the languages Hindi, Telugu and Tamil are given in Table 4.9. Using the corresponding transformations, syllable durations are recomputed from their predicted values. Efficiency of incorporation of the durational constraints and the transformations are analyzed by comparing the performance measures between the transformed duration values and the actual duration values. The performance

Table 4.9 Optimum parameters for the transformations correspond to different categories of syllables in different languages.

Language	Syllable position	a	b	α	β
Hindi	WI	$\mu - 0.6 * \sigma$	$\mu + 0.6 * \sigma$	1.2	1.4
	WM	$\mu - 0.8 * \sigma$	$\mu + 0.8 * \sigma$	1.1	1.2
	WF	$\mu - 0.7 * \sigma$	$\mu + 0.6 * \sigma$	1.2	1.5
	PI	$\mu - 0.6 * \sigma$	$\mu + 0.7 * \sigma$	1.2	1.5
	PM	$\mu - 0.8 * \sigma$	$\mu + 0.9 * \sigma$	1.1	1.2
	PF	$\mu - 0.7 * \sigma$	$\mu + 0.6 * \sigma$	1.3	1.5
Telugu	WI	$\mu - 0.5 * \sigma$	$\mu + 0.7 * \sigma$	1.2	1.6
	WM	$\mu - 0.7 * \sigma$	$\mu + 0.7 * \sigma$	1.2	1.4
	WF	$\mu - 0.5 * \sigma$	$\mu + 0.6 * \sigma$	1.2	1.5
	PI	$\mu - 0.5 * \sigma$	$\mu + 0.6 * \sigma$	1.2	1.5
	PM	$\mu - 0.7 * \sigma$	$\mu + 0.8 * \sigma$	1.2	1.2
	PF	$\mu - 0.5 * \sigma$	$\mu + 0.5 * \sigma$	1.3	1.5
Tamil	WI	$\mu - 0.5 * \sigma$	$\mu + 0.6 * \sigma$	1.2	1.5
	WM	$\mu - 0.6 * \sigma$	$\mu + 0.7 * \sigma$	1.1	1.3
	WF	$\mu - 0.4 * \sigma$	$\mu + 0.6 * \sigma$	1.2	1.6
	PI	$\mu - 0.6 * \sigma$	$\mu + 0.6 * \sigma$	1.3	1.6
	PM	$\mu - 0.7 * \sigma$	$\mu + 0.8 * \sigma$	1.1	1.2
	PF	$\mu - 0.5 * \sigma$	$\mu + 0.6 * \sigma$	1.2	1.6

measures after incorporating the duration constraints are given in Table 4.10. The numbers within brackets indicate the performance measures derived from the transformed (without duration constraints) and actual duration values. The performance has improved compared to the transformation without incorporating the durational constraints.

Table 4.10 Performance measures after imposing the duration constraints. The numbers within brackets indicate the performance measures derived from the transformed (without duration constraints) and actual duration values.

Language	% Transformed syllables within deviation			Objective measures		
	10%	25%	50%	μ (ms)	σ (ms)	γ
Hindi	30(30)	70(69)	88(86)	30(31)	24(25)	.76(.76)
Telugu	30(28)	69(68)	89(87)	27(28)	22(22)	.78(.77)
Tamil	35(34)	77(76)	96(96)	24(26)	22(21)	.83(.82)

4.7 Duration Modeling using Two-Stage Approach

Since a single feedforward neural network model (or SVM model) is used for predicting the durations of syllables for the entire range of 40-300 ms, the accuracy of prediction (Figs. 4.4 and 4.6) is biased towards the mean of the distribution of the training data. This leads to poor prediction for long and short duration syllables which lie at the tail portions of the distributions. This problem can be alleviated to some extent by using multiple models, one for each limited range of duration [68, 135]. This ensures that the training data used for each model is uniformly distributed and well balanced within the limited interval associated with the model. Here the number of models (number of intervals) is not crucial. The optimum number of models and the range of each interval can be arrived at experimentally. But this requires preprocessing of data, which categorizes syllables into different groups based on duration.

For implementing this concept, a two-stage duration model is proposed. The first stage consists of a syllable classifier which groups the syllables based on their duration. The second stage is a function approximator for modeling the syllable duration, which consists of specific models for the given duration interval. The block diagram of the proposed two-stage duration model is shown in Fig. 4.8. The performance of

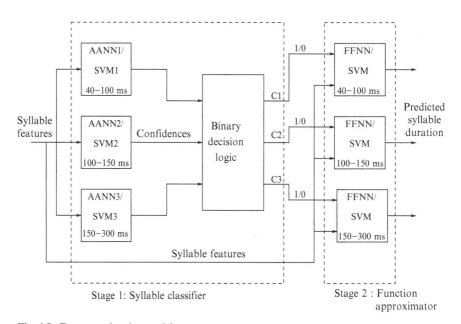

Stage 1: Syllable classifier   Stage 2 : Function
                                                             approximator

Fig. 4.8 Two-stage duration model.

the proposed model depends on the performance of the syllable classifier (1st stage), since the error at the 1st stage will route the syllable features to the unintended model for prediction. The duration intervals arrived at empirically for the language Hindi are 40-120 ms, 120-170 ms and 170-300 ms, whereas for the languages Telugu and Tamil, the intervals are 40-100 ms, 100-150 ms and 150-300 ms.

4.7.1 Syllable classification

In a two-stage duration model, the first stage is the syllable classifier, which classifies the syllables into broad categories according to the chosen duration intervals. Autoassociative Neural Network (AANN) models and Support Vector Machine (SVM) models are explored separately for syllable classification.

A five layer AANN model shown in Fig. 4.9 is used for syllable classification. Autoassociative neural network models are feedforward neural networks performing

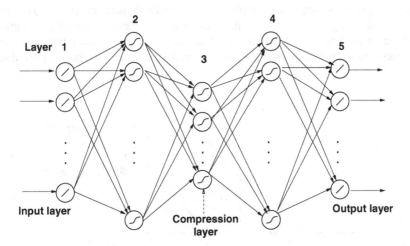

Fig. 4.9 Autoassociative neural network model.

an identity mapping of the input space, and are used to capture the distribution of the input data [100, 136, 137]. The optimum network structures arrived at for the study for Hindi, Telugu and Tamil are 25L 36N 15N 36N 25L, 25L 36N 17N 36N 25L and 25L 36N 12N 36N 25L, respectively. For each language separate AANN models are developed for each of the chosen duration intervals (for Hindi: 40-120, 120-170 and 170-300 ms, and for Telugu and Tamil: 40-100, 100-150 and 150-300 ms). For classification, the syllable features are presented to each of the models. The output of each model is compared with the input to compute the mean squared error. The error (e) is transformed into a confidence (c) value by using the relation $c = exp(-e)$. The confidence values are given to a decision logic, where the highest confidence value among the models is used for classification. The classification performance of the AANN models is shown in Table 4.11.

SVMs are designed for two-class pattern classification. Multiclass (n-class) pattern classification problems can be solved using a combination of binary (2-class) support vector machines. One-against-the-rest approach is used for decomposition of the n-class pattern classification problem into n two-class classification problems. The classification system consists of three SVMs. The set of training examples $\{\{(\mathbf{x}_i,k)\}_{i=1}^{N_k}\}_{k=1}^{n}$ consists of N_k number of examples belonging to the kth class, where the class label $k \in \{1,2,\ldots,n\}$. All the training examples are used to construct the SVM for a class. The SVM for the class k is constructed using the set of training examples and their desired outputs, $\{\{(\mathbf{x}_i,y_i)\}_{i=1}^{N_k}\}_{k=1}^{n}$. The desired output y_i for a training example $\mathbf{x}_i$ is defined as follows:

$$y_i = \begin{cases} +1 & : \quad if \quad \mathbf{x}_i \in k^{th} class \\ -1 & : \quad otherwise \end{cases}$$

The examples with $y_i = +1$ are called positive examples, and those with $y_i = -1$ are called negative examples. An optimal hyperplane is constructed to separate positive examples from negative examples. The separating hyperplane (margin) is chosen in such a way as to maximize its distance from the closest training examples of different classes [100, 130]. Fig. 4.10 illustrates the geometric construction of a hyperplane for two dimensional input space. The support vectors are those data

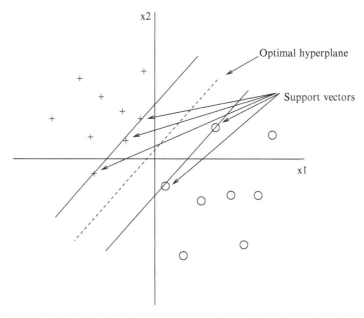

Fig. 4.10 Illustration of the idea of support vectors and an optimum hyper plane for linearly separable patterns.

points that lie closest to the decision surface, and therefore the most difficult to classify. They have a direct bearing on the optimum location of the decision surface.

For a given test pattern $\mathbf{x}$, the evidence $D_k(\mathbf{x})$ is obtained from each of the SVMs. In the decision logic, the class label k associated with the SVM which gives maximum evidence is hypothesized as the class (C) of the test pattern. That is,

$$C(\mathbf{x}) = \arg\max_{k} \ (D_k(\mathbf{x}))$$

The decision logic provides three outputs (corresponding to the number of duration intervals), which are connected to the corresponding models in the following stage (function approximator). For each syllable the decision logic activates only the duration model corresponding to the maximum confidence. The selected duration model now predicts the duration of the syllable in that limited interval. The performance of classification using SVM models is shown in Table 4.11. The SVM model seems to be better than AANN model for the 1st stage classification.

Table 4.11 Classification performance using AANN models and SVMs for the languages Hindi, Telugu and Tamil.

Language	% syllables correctly classified	
	AANN models	SVM models
Hindi	74.68	81.92
Telugu	79.22	80.17
Tamil	76.17	83.26

4.7.2 Evaluation of two-stage duration model

For modeling the durations of syllables using the two-stage model, syllable features are presented to all the models of the syllable classifier. The decision logic in the syllable classifier routes the syllable features to one of the duration models present in the second stage (function approximator) for predicting the duration of the syllable in a specific limited duration interval. The performance of the proposed two-stage duration model is given in Table 4.12. The numbers within brackets indicate the performance of the single FFNN model. In this study SVM models are used in syllable classification stage (1st stage), and FFNN models are used in function approximator stage (2nd stage). A similar performance is observed by using SVM models in function approximator stage.

In the proposed approach, the duration models are developed with nonoverlapped intervals (for Hindi: 40-120, 120-170 and 170-300, for Telugu and Tamil: 40-100, 100-150 and 150-300 ms). Due to this, the syllables having durations around 120 and 170 ms (lying on the border of the two intervals) in Hindi and syllables with durations 100 and 150 ms in Telugu and Tamil, may be affected, leading to their poor prediction. This can be reduced by developing the duration models with overlapped

Table 4.12 Performance of the two-stage model using nonoverlapped intervals. The numbers within brackets indicate the performance of the single FFNN model.

Language	% Predicted syllables within deviation			Objective measures		
	10%	25%	50%	μ (ms)	σ (ms)	γ
Hindi	36(29)	79(68)	96(84)	26(32)	20(26)	.81(.75)
Telugu	38(29)	81(66)	95(86)	23(29)	23(23)	.82(.78)
Tamil	44(34)	85(75)	97(96)	21(26)	21(22)	.84(.82)

duration intervals. Therefore models are developed with the following duration intervals: 40-140, 100-190, 150-300 ms for Hindi and 40-120, 80-170 and 130-300 ms for Telugu and Tamil. The performance of the two-stage model with the overlapped duration intervals is given in Table 4.13. The numbers within brackets indicate the performance of the two-stage model with nonoverlapped intervals.

Table 4.13 Performance of the two-stage model using overlapped intervals. The numbers within brackets indicate the performance of the two-stage model with nonoverlapped intervals.

Language	% Predicted syllables within deviation			Objective measures		
	10%	25%	50%	μ (ms)	σ (ms)	γ
Hindi	37(36)	80(79)	96(96)	25(26)	20(20)	.82(.81)
Telugu	39(38)	83(81)	96(95)	23(23)	23(23)	.82(.82)
Tamil	44(44)	86(85)	97(97)	20(21)	20(21)	.85(.84)

Results show that the prediction accuracy is improved with two-stage model compared to single model for the entire duration range. For comparing the performance of two-stage model with single FFNN model, Tamil broadcast news data is chosen. The prediction performance of single FFNN model and two-stage model are shown in Fig. 4.11. Performance curves in the figure show that the syllables having duration around the mean of the distribution are estimated better in both the models, whereas the short and long duration syllables are poorly predicted in the case of single FFNN model. The prediction accuracy of these extreme (long and short) syllables is improved in the two-stage model, because of use of specific models for each duration interval.

The prediction performance of the proposed models can be compared with the results using Classification and Regression Tree (CART) models. The performance of CART models is given in Table 4.14. The performance of FFNN models (shown within brackets in Table 4.14) is comparable to that of CART models, whereas the two-stage models (Table 4.13) seem to perform better than the CART models.

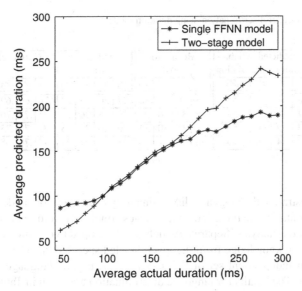

Fig. 4.11 Prediction performance of single FFNN model and two-stage model

Table 4.14 Performance of the CART models. The numbers within brackets indicate the performance of FFNN models.

Language	% Predicted syllables within deviation			Objective measures		
	10%	25%	50%	μ (ms)	σ (ms)	γ
Hindi	31(29)	67(68)	92(84)	32(32)	26(26)	.76(.75)
Telugu	30(29)	64(66)	88(86)	29(29)	24(23)	.78(.78)
Tamil	33(34)	71(75)	93(96)	25(26)	21(22)	.81(.82)

4.8 Duration Models for Speaker and Language Identification

For demonstrating the speaker discriminability by the duration models, speech data of four different speakers of Hindi is considered. Separate models are developed to capture the speaker-specific duration knowledge from the given labeled speech data. The speaker models are denoted by $\mathscr{S}_1, \mathscr{S}_2, \mathscr{S}_3$ and $\mathscr{S}_4$. The models were tested with an independent set of data for determining the speaker discrimination. The results of the speaker discrimination are given in Table 4.15. The results show that the percentage of syllables predicted within 25% deviation from the actual duration is highest for the intended speaker compared to others. Likewise the average prediction error (μ), standard deviation (σ) values are lowest, and the correlation coefficient (γ) values are highest for the intended speaker.

Table 4.15 Speaker discrimination using duration models.

	% predicted syllables within 25% deviation				Mean error (ms)				Std. dev. (ms)				Corr. coeff.			
	$\mathscr{S}_1$	$\mathscr{S}_2$	$\mathscr{S}_3$	$\mathscr{S}_4$	$\mathscr{S}_1$	$\mathscr{S}_2$	$\mathscr{S}_3$	$\mathscr{S}_4$	$\mathscr{S}_1$	$\mathscr{S}_2$	$\mathscr{S}_3$	$\mathscr{S}_4$	$\mathscr{S}_1$	$\mathscr{S}_2$	$\mathscr{S}_3$	$\mathscr{S}_4$
$\mathscr{S}_1$	58	51	53	55	33	38	37	36	27	30	29	29	.76	.69	.69	.71
$\mathscr{S}_2$	54	56	52	51	35	32	36	37	29	24	26	27	.71	.73	.70	.69
$\mathscr{S}_3$	53	52	60	52	36	35	30	34	30	26	23	26	.72	.71	.75	.71
$\mathscr{S}_4$	53	54	56	59	36	35	34	30	28	27	27	24	.73	.70	.68	.76

For demonstrating the language discriminability using duration models, a labeled set of speech data for the three Indian languages Hindi (Hi), Telugu (Te), and Tamil (Ta) is chosen for analysis. Separate models are developed to capture the language-specific duration knowledge from the given labeled speech data. The models were tested with an independent set of the data for showing the language discrimination capability. The results for language discrimination are given in Table 4.16. The results show that the duration models also capture the language-specific information. Hence this ability of the duration models to capture the speaker-specific and language-specific information can be exploited in speaker and language identification studies.

Table 4.16 Language discrimination using duration models.

	% predicted syllables within 25% deviation			Mean error (ms)			Std. dev. (ms)			Corr. coeff.		
	Hi	Te	Ta	Hi	Te	Ta	Hi	Te	Ta	Hi	Te	Ta
Hi	68	59	57	32	37	38	24	28	30	0.76	0.71	0.69
Te	51	66	58	37	29	32	27	23	25	0.72	0.79	0.73
Ta	55	61	74	35	33	26	27	26	22	0.71	0.75	0.83

4.9 Summary

Feedforward neural network models and support vector machine models were proposed for predicting the durations of syllables in the languages Hindi, Telugu and Tamil [138–141]. The duration of the sound unit depend on its linguistic context and production constraints. These were represented with positional, contextual and phonological features, and used for developing the models. The effect of duration and intonation constraints was examined by modeling the durations of the syllables with these constraints as input to the models [142]. The prediction performance of the models is improved by considering these constraints for developing the models. Processing the predicted durations further using piecewise linear transformation, improves the accuracy of prediction. A two-stage duration model was proposed to reduce the problem of poor prediction due to a single FFNN (or SVM) model [143]. The performance of the two-stage model was improved by appropriate syllable classification model and the selection criterion of duration intervals. The performance of the neural network models is compared with the results obtained by CART models. The ability of the duration models to capture speaker-specific and language-specific information was also demonstrated by developing individual models for different speakers and languages [144]. This information is useful as evidence in speaker and language identification tasks. In the next chapter FFNN and SVM models are developed for predicting the intonation patterns for the sequence of syllables.

Chapter 5
MODELING INTONATION

Abstract This chapter discuss about modeling of intonation patterns using feedforward neural networks (FFNN) and Support Vector Machines (SVMs). Linguistic and production constraints are represented in the form of positional, contextual and phonological features for modeling the intonation patterns of the sequence of syllables. Prediction accuracy of the proposed models is analyzed using objective measures, and also compared with Classification and Regression Tree (CART) models. Postprocessing methods using intonation constraints are proposed for improving the accuracy of prediction. The ability of the intonation models to capture the language specific and speaker-specific information is demonstrated.

5.1 Introduction

In the previous chapter feedforward neural networks and support vector machines are explored for capturing the implicit duration knowledge present in the speech. It was observed that the duration models also capture speaker-specific and language-specific information. In this chapter feedforward neural networks and support vector machines are proposed for capturing the intonation patterns of speech in Indian languages.

Intonation knowledge is not explicitly taught or learned when we learn to speak. Hence it is difficult to state the rules governing the intonation patterns for an utterance. Extracting the implicit rules corresponding to the models of the intonation patterns of the sound units present in the speech signal is a difficult task. In a speech signal, the intonation pattern (F_0 contour) corresponding to a sequence of sound units is constrained by the linguistic context of the units and the constraints in the production of those units. Hence for modeling the intonation, features representing the linguistic and production constraints of the sound units are used. There are two objectives in this study. The first one is to determine whether nonlinear models can

K.S. Rao, *Predicting Prosody from Text for Text-to-Speech Synthesis*, SpringerBriefs   65
in Electrical and Computer Engineering, DOI 10.1007/978-1-4614-1338-7_5,
© Springer Science+Business Media New York 2012

capture the implicit knowledge of the intonation patterns of syllables in a language. The ability of the models to capture the intonation knowledge is examined by using the speech spoken by various speakers in different Indian languages. The second objective is to demonstrate that the intonation models can capture the speaker-specific and language-specific information, and this in turn may be used for speaker and language identification studies.

This chapter is organized as follows: The features used as input to the neural network or SVM for capturing the knowledge of intonation patterns for the sequence of syllables are described in Section 5.2. The database used for intonation modeling is described in Section 5.3. The performance of the neural network model for predicting the intonation patterns is given in Section 5.4. Section 5.5 discusses intonation modeling using support vector machines. Postprocessing of the predicted F_0 values using the knowledge of intonation constraints is described in Section 5.6. Section 5.7 demonstrates the ability of the intonation models for identifying the speaker and language from the given speech.

5.2 Features for Developing the Intonation Model

The features used for modeling the intonation patterns for the sequence of syllables are almost same as those used for modeling the durations of syllables. Details of these features were discussed in Chapter 4. The list of features and the number of input nodes in a neural network needed to represent the features are given in Table 5.1.

Table 5.1 List of factors affecting the F_0 of a syllable, features representing the factors and the number of nodes needed for neural network to represent the features.

Factors	Features	# Nodes
Syllable position in the phrase	Position of syllable from beginning of the phrase Position of syllable from end of the phrase Number of syllables in the phrase	3
Syllable position in the word	Position of syllable from beginning of the word Position of syllable from end of the word Number of syllables in the word	3
Word position in the phrase	Position of word from beginning of the phrase Position of word from end of the phrase Number of words in the phrase	3
Syllable identity	Segments of the syllable (consonants and vowels)	4
Context of the syllable	Previous syllable	4
	Following syllable	4
Syllable nucleus	Position of the nucleus Number of segments before the nucleus Number of segments after the nucleus	3
Pitch	F_0 of the previous syllable	1

5.3 Database for Intonation Modeling

The database used for modeling the intonation patterns is same as that used for modeling the durations of syllables. The detailed description of the database was given in Chapter 4. For developing intonation models, the fundamental frequencies (F_0) of the syllables should be available in the database. The fundamental frequencies of the syllables are computed using the autocorrelation of the Hilbert envelope of the linear prediction residual [145]. The average pitch (F_0) for male speakers and female speakers in the database was found to be 129 and 231 Hz, respectively.

5.4 Intonation Modeling using Feedforward Neural Networks

A four layer feedforward neural network is used for modeling the intonation patterns for the sequence of syllables. The details of the architecture and function of the network were explained in Chapter 4. A separate model is developed for each speaker in the database in the three languages. The extracted input vectors representing positional, contextual and phonological features are presented as input, and the corresponding F_0 values of the syllables are presented as desired outputs to the FFNN model. The network is trained for 500 epochs. Training errors for neural network models for one speaker in each of the three languages are shown in Fig. 5.1. The prediction performance of the models for one female speaker for each of the three languages is shown in Fig. 5.2. Each plot represents the average predicted F_0 vs the average F_0 of a syllable. The prediction performance of the intonation models is illustrated in Table 5.2 for one male (M) and one female (F) speaker for each language.

Table 5.2 Performance of the FFNN models for predicting the F_0 values for different languages. The results for female (F) and male (M) speakers are given separately.

Language # syllables	Gender	% Predicted syllables within deviation			Objective measures		
		10%	15%	25%	μ (Hz)	σ (Hz)	γ
Hindi	F(660)	67	82	96	20	18	0.78
	M(1143)	74	92	98	12	9	0.79
Telugu	F(1276)	72	91	99	16	13	0.78
	M(984)	64	82	96	10	9	0.79
Tamil	F(741)	77	91	99	18	14	0.85
	M(1267)	77	90	97	13	12	0.80

Fig. 5.3 shows the predicted and actual (original) pitch contours for the utterance *"pradhAn mantri ne kahA ki niyantran rekhA se lekar"* in Hindi spoken by a male speaker. It shows that the predicted pitch contour is close to the original contour.

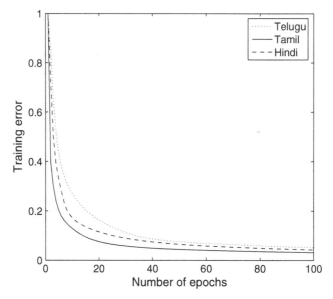

Fig. 5.1 Training errors of the neural network models for predicting the intonation patterns for the sequence of syllables in three languages Hindi, Telugu and Tamil.

This indicates that the neural network predicts the F_0 values reasonably well for the sequence of syllables in the given text.

The prediction performance of the neural network models can be compared with the results obtained by Classification and Regression Tree (CART) models. The performance of the CART models is given in Table 5.3. The performance of FFNN models (shown within brackets in Table 5.3) seems to be better than for the CART models.

Table 5.3 Performance of the CART models for predicting the F_0 values for different languages. The numbers within brackets indicate the performance of neural network models.

Lang-uage	Gender	% Predicted syllables with deviation			Objective measures		
		10%	15%	25%	μ(Hz)	σ(Hz)	γ
Hindi	F	58(67)	73(82)	90(96)	21(20)	20(18)	.76(.78)
	M	68(74)	90(92)	98(98)	14(12)	11(9)	.77(.79)
Telugu	F	68(72)	87(91)	99(99)	18(16)	13(13)	.74(.78)
	M	64(64)	83(82)	96(96)	11(10)	11(9)	.74(.79)
Tamil	F	69(77)	90(91)	99(99)	21(18)	15(14)	.80(.85)
	M	72(77)	88(90)	95(97)	15(13)	13(12)	.77(.80)

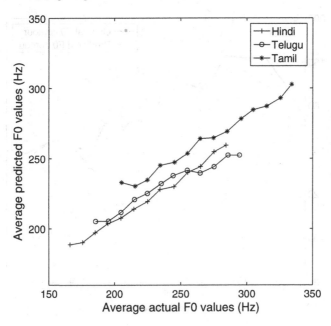

Fig. 5.2 Prediction performance of the neural network model for female speakers in the languages Hindi, Telugu and Tamil.

For studying the impact of duration and intonation constraints in predicting the intonation patterns of the syllables, separate models are developed using different constraints. They are: (a) Model with linguistic features, (b) model with linguistic features and duration constraints, (c) model with linguistic features and intonation constraints and (d) model with linguistic features, duration and intonation constraints. The performance of these models is given in Table 5.4. The results indicate that the prediction performance has improved by imposing the constraints. Better performance is observed when all the constraints are applied together.

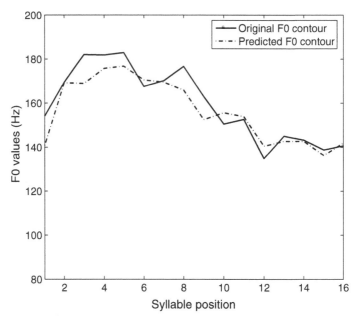

Fig. 5.3 Comparison of predicted F_0 contour with original contour for the utterance "*pradhAn mantri ne kahA ki niyantran rekhA se lekar*" in Hindi spoken by a male speaker.

Table 5.4 Performance of the neural network models for prediction of the F_0 values of the syllables using different constraints.

Features	Lang. gender	% Predicted syllables within deviation			Objective measures		
		10%	15%	25%	μ (Hz)	σ (Hz)	γ
Linguistic	Hi(M)	63	80	98	15	12	0.68
	Te(F)	64	82	98	20	16	0.71
	Ta(F)	67	87	99	21	17	0.74
Linguistic and duration	Hi(M)	71	87	98	13	11	0.72
	Te(F)	74	89	99	17	15	0.73
	Ta(F)	73	90	98	19	17	0.77
Linguistic and intonation	Hi(M)	78	91	99	11	11	0.76
	Te(F)	79	93	99	16	13	0.78
	Ta(F)	77	93	99	17	14	0.81
Linguistic, duration and intonation	Hi(M)	82	93	99	9	8	0.80
	Te(F)	81	94	100	16	12	0.81
	Ta(F)	84	95	99	16	13	0.84

5.5 Intonation Modeling using Support Vector Machines

Support vector machines predict the F_0 value of a syllable using regression (function approximation). The architecture of the SVM model for regression, and the basic function of SVM were discussed in Chapter 4. Training and testing data used for evaluating the performance of the SVM regression model is the same as those used for neural network models discussed in Section 5.4. The performance of the SVM regression model in predicting the F_0 of the syllables for two speakers (male (M) and female (F)) in each language is given in Table 5.5.

Table 5.5 Performance of the SVM models for predicting the F_0 values for different languages. The results for female (F) and male (M) speakers are given separately.

Language	Gender	% Predicted syllables within deviation			Objective measures		
		10%	15%	25%	μ (Hz)	σ (Hz)	γ
Hindi	F	63	76	93	19	19	0.80
	M	76	91	98	12	10	0.78
Telugu	F	76	92	99	15	13	0.79
	M	68	85	97	10	10	0.78
Tamil	F	76	92	99	18	14	0.83
	M	75	91	98	13	12	0.81

The prediction performance of one of the models of the female speakers in each language is shown in Fig. 5.4. The effect of intonation and duration constraints for predicting the F_0 values of the syllables using SVM models is similar to FFNN models with these constraints.

The performance measures indicate that SVM models capture the intonation patterns similar to neural network models. The overall prediction error of the models is about 12 Hz (9.3% w.r.t mean F_0) for male speakers and 19 Hz (8.2% w.r.t mean F_0) for female speakers. About 87% of the syllables are predicted within 15% deviation.

5.6 Postprocessing of the Predicted F_0 Values

The accuracy of prediction (Figs. 5.2 and 5.4) of the intonation models is not uniform for the entire range of F_0 values. The accuracy of prediction can be improved to some extent by imposing the piecewise linear transformation on the output predicted F_0 values [132–134]. The details of the piecewise linear transformation were discussed in Chapter 4.

The performance measures derived from the transformed and actual F_0 values are given in Table 5.6. The numbers within brackets indicate the performance measures derived from the predicted (without transformation) and actual F_0 values. The

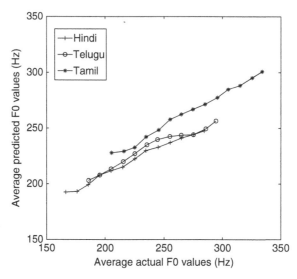

Fig. 5.4 Prediction performance of SVM model for female speakers in the languages Hindi, Telugu and Tamil.

prediction performance has improved slightly after imposing the transformation. The performance may be further improved by incorporating the intonation constraints over the transformed F_0 values.

Table 5.6 Performance measures after imposing the transformation. The numbers within brackets indicate the performance measures derived from the predicted (without transformation) and actual F_0 values.

Lang-uage	Gender	% Predicted syllables within deviation			Objective measures		
		10%	15%	25%	μ(Hz)	σ(Hz)	γ
Hindi	F	66(67)	84(82)	99(96)	20(20)	18(19)	.79(.78)
	M	75(74)	93(92)	98(98)	12(13)	10(9)	.80(.79)
Telugu	F	73(72)	91(91)	98(99)	16(17)	13(13)	.78(.78)
	M	63(64)	83(82)	97(96)	10(10)	10(9)	.80(.79)
Tamil	F	76(77)	93(91)	99(99)	18(18)	14(14)	.84(.85)
	M	78(77)	92(90)	97(97)	13(14)	12(12)	.81(.80)

The intonation patterns (sequence of F_0 values) for the sequence of syllables will follow some constraints at the word level and phrase level. The F_0 range (difference between the maximum and minimum values of F_0 for the syllables in a word) of a word depends on the number of syllables present in the word, and on the position of the word in the phrase. According to size (in number of syllables) of the word, the F_0 range is analyzed with respect to bisyllabic, trisyllabic, tetrasyllabic and polysyllabic words. As per the position of a word in the phrase, the F_0 range is analyzed with respect to the first, second, third, fourth, fifth, sixth, and other higher positions in the phrase. The constraints at the phrase level include the F_0 range across the phrase. The change in F_0 (F_0 resetting) from the end of a word to the beginning of the next word, and the end of the previous phrase to the beginning of the present phrase are also considered as constraints for adjusting the transformed F_0 values to reduce the error further. The intonation constraints (F_0 range and F_0 resetting) are estimated for different speakers in the languages Hindi, Telugu and Tamil. The constraints associated with F_0 range values at the word and phrase levels for the broadcast news read by a male news reader in Hindi are given in Table 5.7. The constraints of F_0 resetting at the word and phrase levels are given in Table 5.8. The transformed F_0 values

Table 5.7 Intonation constraints associated with F_0 range at word level and phrase level for the broadcast news data read by a male news reader in Hindi.

Word/Phrase	Word type	F_0 range (Hz)	
		Mean (μ)	Std. dev. (σ)
Size of a word	Bisyllabic	11.58	9.94
	Trisyllabic	18.07	11.6
	Tetrasyllabic	23.3	12.2
	Polysyllabic	31.35	13.73
Position of a word	First	17.49	12.7
	Second	15.69	11.55
	Third	13.77	11.31
	Fourth	13.35	10.87
	Fifth	12.69	10.57
	Sixth	12.17	10.42
	> Sixth	12.11	10.31
Phrase		50.14	19.97

are adjusted by imposing the intonation constraints. The effectiveness of the intonation constraints is examined by comparing the performance measures before and after imposing the constraints. The performance measures after incorporating the intonation constraints are given in Table 5.9. The performance has improved when compared to the transformation without incorporating the intonation constraints.

5.7 Intonation Models for Speaker and Language Identification

For demonstrating speaker discriminability using intonation models, speech data of four different female speakers of Hindi language is considered. Separate models

Table 5.8 Intonation constraints associated with F_0 resetting at word level and phrase level for the broadcast news data read by a male news reader in Hindi.

Word/	Word type	F_0 change (Hz)	
Phrase		Mean (μ)	Std. dev. (σ)
Word	First and second	15.36	13.14
level	Second and third	14.12	12.83
	Third and fourth	13.12	12.43
	Fourth and fifth	12.71	11.07
	Fifth and sixth	12.06	10.55
	Sixth and seventh	11.66	10.21
Phrase		21.48	17.21

Table 5.9 Performance measures derived after imposing the intonation constraints over transformed values. The numbers within brackets indicate the performance measures derived from the transformed (without imposing intonation constraints) and actual F_0 values.

Lang-uage	Gender	% Predicted syllables within deviation			Objective measures		
		10%	15%	25%	μ(Hz)	σ(Hz)	γ
Hindi	F	68(66)	85(84)	99(99)	19(20)	18(18)	.79(.79)
	M	77(75)	93(93)	98(98)	11(12)	9(9)	.81(.80)
Telugu	F	74(73)	93(91)	99(98)	16(16)	12(13)	.79(.78)
	M	66(63)	87(83)	99(97)	9(10)	9(9)	.81(.80)
Tamil	F	78(76)	93(93)	99(99)	17(18)	14(14)	.85(.84)
	M	78(78)	93(92)	98(97)	12(13)	11(12)	.81(.81)

are developed to capture the speaker-specific intonation knowledge from the given labeled speech data. The speaker models are denoted by $\mathscr{S}_1, \mathscr{S}_2, \mathscr{S}_3$ and $\mathscr{S}_4$. The results of the speaker discrimination are given in Table 5.10. The results show that

Table 5.10 Speaker discrimination using intonation models.

	% predicted syllables within 15% deviation				Mean error (Hz)				Std. dev. (Hz)				Corr. coeff.			
	$\mathscr{S}_1$	$\mathscr{S}_2$	$\mathscr{S}_3$	$\mathscr{S}_4$	$\mathscr{S}_1$	$\mathscr{S}_2$	$\mathscr{S}_3$	$\mathscr{S}_4$	$\mathscr{S}_1$	$\mathscr{S}_2$	$\mathscr{S}_3$	$\mathscr{S}_4$	$\mathscr{S}_1$	$\mathscr{S}_2$	$\mathscr{S}_3$	$\mathscr{S}_4$
$\mathscr{S}_1$	92	65	83	81	16	28	24	21	14	19	18	19	.79	.73	.74	.75
$\mathscr{S}_2$	75	92	66	69	27	18	31	26	23	14	24	23	.77	.83	.73	.74
$\mathscr{S}_3$	68	71	85	76	25	26	19	24	20	27	15	19	.72	.75	.82	.76
$\mathscr{S}_4$	72	80	79	88	25	23	24	18	21	19	18	15	.76	.77	.78	.82

the percentage of syllables predicted within 15% deviation from their actual F_0 values is highest for the intended speaker compared to others. Likewise the average prediction error (μ), standard deviation (σ) values are lowest, and the correlation coefficient (γ) values are highest for the intended speaker.

For demonstrating the language discriminability using intonation models, speech data of male speakers from Hindi (Hi), Telugu (Te), and Tamil (Ta) is chosen for analysis. Separate models are developed to capture the language-specific intonation knowledge from the given labeled speech data. The results for language discrimination are given in Table 5.11. The results show that the intonation models also capture the language-specific information.

Table 5.11 Language discrimination using intonation models.

	% predicted syllables within 15% deviation			Mean error (Hz)			Std. dev. (Hz)			Corr. coeff.		
	Hi	Te	Ta	Hi	Te	Ta	Hi	Te	Ta	Hi	Te	Ta
Hi	90	59	63	11	22	20	9	14	14	0.79	0.75	0.68
Te	42	85	61	20	10	16	13	10	12	0.74	0.78	0.73
Ta	56	65	81	19	17	13	14	13	12	0.74	0.76	0.80

5.8 Summary

Feedforward neural network and support vector machine models were proposed for predicting the F_0 values of the syllables in the languages Hindi, Telugu and Tamil [141, 146, 147]. F_0 range and F_0 resetting at word level and phrase level were used as constraints for modifying the predicted values. The prediction performance was shown to improve by processing the predicted values using piecewise linear transformation and imposing the intonation constraints. The intonation models also capture the speaker and language information besides the intonation knowledge [142, 148–150]. This was verified with the models of different speakers and different languages. The speaker-specific and language-specific knowledge may be useful for speaker and language identification tasks [144].

Chapter 6
PROSODY MODIFICATION

Abstract This chapter focuses on the incorporation of duration and intonation information in a speech signal. A new method is proposed for the modification of pitch and duration of a given speech signal using the instants of significant excitation as anchor points. The method is based on exploiting the nature of excitation of speech production mechanism. The performance of the proposed method for prosody modification is compared with some standard methods.

6.1 Introduction

Previous chapters discussed on prosody models using neural networks and support vector machines for capturing the implicit duration and intonation knowledge present in a speech signal. To incorporate the acquired prosody knowledge, a new method of prosody modification of speech is proposed in this chapter. The method is based on exploiting the nature of excitation of speech production mechanism.

The objective of prosody modification is to alter the pitch contour and durations of the sound units of speech without affecting the shapes of the short-time spectral envelopes[72]. A method for prosody (pitch and duration) modification is proposed using the knowledge of the instants of significant excitation. The instants of significant excitation refer to the instants of glottal closure in the voiced region, and to some random excitations like the onset of burst in the case of nonvoiced regions [151, 152]. The instants of significant excitation are also termed as *epochs*. The proposed method does not distinguish between voiced and nonvoiced regions in the implementation of the desired prosody modification. The method also does not involve estimation of any specific speech parameters like fundamental frequency (F_0). Since the modification is done in the excitation component of the signal, there will be no perceived discontinuities in the synthesized speech even when large values of prosody modification factors are used.

K.S. Rao, *Predicting Prosody from Text for Text-to-Speech Synthesis*, SpringerBriefs
in Electrical and Computer Engineering, DOI 10.1007/978-1-4614-1338-7_6,
© Springer Science+Business Media New York 2012

Methods for prosody modification generally produce some spectral and phase distortions. This is mainly due to manipulation of the speech signal directly. The distortions are reduced to a large extent in the proposed method by operating on the residual obtained from the linear prediction analysis. In this work we propose a method for prosody modification, which operates on the linear prediction residual using the knowledge of the instants of significant excitation as pitch markers. The instants of significant excitation are computed using group-delay analysis [152]. The group-delay-based method is robust, and it gives accurate epoch locations even under some mild degradations due to background noise and reverberation [151]. The region around the instant of glottal closure correspond to the significant part of excitation, in the sense that the strength of excitation is maximum in that region of the pitch period. Therefore, we attempt to retain that region during pitch period modification. Since methods like LP-PSOLA also use residual manipulation for prosody modification, we compare the performance of our method with the results obtained by LP-PSOLA. An important feature of the proposed method is that the instants of significant excitation in both the voiced and nonvoiced regions are treated alike.

This chapter is organized as follows: The basic principle of the proposed method for prosody modification is presented in Section 6.2. Modification of pitch period is discussed in Section 6.3 and the modification of duration in Section 6.4. After obtaining the modified (new) epoch sequence according to the desired prosody information, the next step is to modify the LP residual according to the new sequence of epochs. The process of modification of the LP residual is discussed in Section 6.5, and the process of generating the synthesized speech by exciting the time varying all-pole filter with the modified residual is discussed in Section 6.6. The performance of the proposed prosody manipulation method is compared with the LP-PSOLA method in Section 6.7. A summary of this chapter is given in Section 6.8.

6.2 Proposed Method for Prosody Modification

The proposed method for prosody manipulation makes use of the properties of the excitation source information for prosody modification. The residual signal in the Linear Prediction (LP) analysis is used as an excitation signal [153]. The successive samples in the LP residual are less correlated compared to the samples in the speech signal. The residual signal is manipulated by using a resampling technique either for increasing or decreasing the number of samples required for the desired prosody modification. The residual manipulation is likely to introduce less distortion in the speech signal synthesized using the modified LP residual and LP coefficients (LPCs). The time varying vocal tract system characteristics are represented by the LPCs for each analysis frame. Since the LPCs carry the information about the short-time spectral envelope, they are not altered in the proposed method for prosody modification. LP analysis is carried out over short segments (analysis frames) of speech data to derive the LP coefficients and the LP residual for the speech signal [153].

There are four main steps involved in the prosody manipulation: (1) Deriving the instants of significant excitation (epochs) from the LP residual signal, (2) deriving a modified (new) epoch sequence according to the desired prosody (pitch and duration), (3) deriving a modified LP residual signal from the modified epoch sequence, and (4) synthesizing speech using the modified LP residual and the LPCs.

In this section we will briefly discuss the method of extracting the instants of significant excitation (or epochs) from the LP residual [151, 152]. Methods for pitch period modification and duration modification are described in Sections 6.3 and 6.4, respectively. Group-delay analysis is used to derive the instants of significant excitation from the LP residual [151, 152]. The analysis involves computation of the average slope of the unwrapped phase spectrum (i.e., average group-delay) for each frame. If $X(\omega)$ and $Y(\omega)$ are the Fourier transforms of the windowed signal $x(n)$ and $nx(n)$, respectively, then the group-delay function $\tau(\omega)$ is given by the negative derivative of the phase function $\phi(\omega)$ of $X(\omega)$, and is given by [151, 154]

$$\tau(\omega) = -\phi'(\omega) = \frac{X_R Y_R + X_I Y_I}{X_R^2 + X_I^2},$$

where $X_R + jX_I = X(\omega)$, and $Y_R + jY_I = Y(\omega)$. Any isolated sharp peaks in $\tau(\omega)$ are removed by using a 3-point median filtering. Note that all the Fourier transforms are implemented using the discrete Fourier transform. The average value $\bar{\tau}$ of the smoothed $\tau(\omega)$ is the value of the *phase slope function* for the time instant corresponding to the center of the windowed signal $x(n)$. The phase slope function is computed by shifting the analysis window by one sample at a time. The instants of positive zero-crossings of the phase slope function correspond to the instants of significant excitation. Figs. 6.1 and 6.2 illustrate the results of extraction of the instants of significant excitation for voiced and nonvoiced speech segments, respectively.

For generating these figures, a 10th order LP analysis is used with a frame size of 20 ms and a frame shift of 5 ms. Throughout this study the signal sampled at 8 kHz is used. The signal in the analysis frame is multiplied with a Hamming window to generate a windowed signal. The algorithm for determining the instants of significant excitation for speech signals using group delay function is given in Table 6.1. Note that for nonvoiced speech, the epochs occur at random instants, whereas for voiced speech the epochs occur in the regions of the glottal closure, where the LP residual error is large. The time interval between two successive epochs correspond to the pitch period for voiced speech. With each epoch we associate three parameters, namely, time instant, epoch interval and LP residual. We call these as *epoch parameters*.

The prosody manipulation involves deriving a new excitation (LP residual) signal by incorporating the desired modification in the duration and pitch period for the utterance. This is done by first creating a new sequence of epochs from the original sequence of epochs. For this purpose all the epochs derived from the original signal are considered, irrespective of whether they correspond to a voiced segment or a nonvoiced segment. The methods for creating the new epoch sequence for the desired prosody modification are discussed in Sections 6.3 and 6.4.

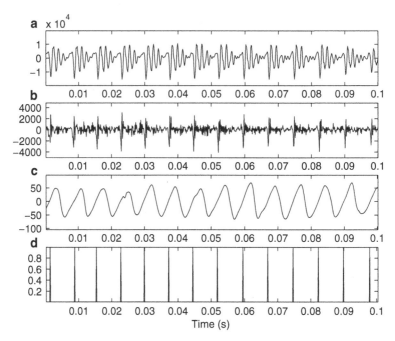

Fig. 6.1 (a) A segment of voiced speech and its (b) LP residual, (c) phase slope function, and (d) instants of significant excitation.

For each epoch in the new epoch sequence, the nearest epoch in the original epoch sequence is determined, and thus the corresponding epoch parameters are identified. The original LP residual is modified in the epoch intervals of the new epoch sequence, and thus a modified excitation (LP residual) signal is generated. The modified LP residual signal is then used to excite the time varying all-pole filter represented by the LPCs. For pitch period modification, the filter parameters (LPCs) are updated according to the frame shift used for analysis of the original signal. For duration modification, the LPCs are updated according to the modified frame shift value. Generation of the modified LP residual according to the desired pitch period and duration modification factors is described in Section 6.5, and the speech synthesis procedure in Section 6.6. Fig. 6.3 shows the block diagram indicating various stages in prosody modification.

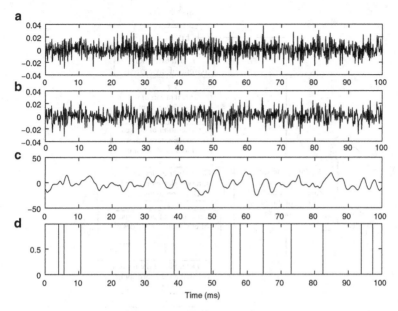

Fig. 6.2 (a) A segment of nonvoiced speech and its (b) LP residual, (c) phase slope function, and (d) instants of significant excitation.

Table 6.1 Steps for determining the instants of significant excitation using group delay function.

1. Preemphasize the speech signal.
2. Compute the LP residual using a frame size of 20 ms, frame shift of 5 ms and 10th order LP analysis.
3. Compute the group delay $(-\phi'(w))$ for each frame of the residual signal of size 10 ms and frame shift of one sample using the relation
$$-\phi'(\omega) = \tau(\omega) = \frac{X_R Y_R + X_I Y_I}{X_R^2 + X_I^2}$$
where $X_R + jX_I = X(\omega)$ and $Y_R + jY_I = Y(\omega)$
$X(\omega)$ is the Fourier transform of $x(n)$ whose group delay is required and $Y(\omega)$ is the Fourier transform of $nx(n)$.
$n = 0, 1, 2, \cdots, N$ and N is length of $x(n)$.
4. Smooth the group delay function using a median filter of order 3 for removing the unwanted peaks.
5. Find the average group delay for each frame which is the required phase slope function.
6. Smooth the phase slope function with a Hamming window of order 8.
7. Identify the positive zero crossings as the instants of significant excitation.

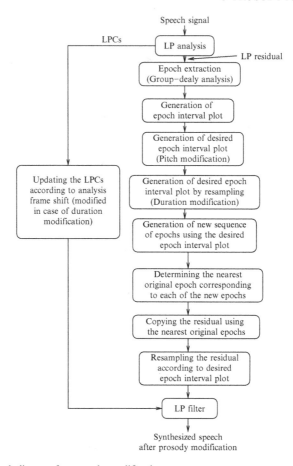

Fig. 6.3 Block diagram for prosody modification.

6.3 Pitch Period Modification

The objective is to generate a new epoch sequence, and then a new LP residual signal according to the desired pitch period modification factor. Note that for generating this signal, all epochs in the sequence are considered, without discriminating the voiced or nonvoiced nature of the segment to which the epoch belongs. Thus it is not necessary to identify the voiced, unvoiced and silence regions of speech.

Fig. 6.4 illustrates the prosody modification where the pitch period is reduced by a factor $\alpha = 0.75$. The original epoch interval plot is shown by the solid curve, and the desired epoch interval plot is shown by the dotted curve, which is obtained by multiplying the solid curve by α. The original epochs are marked by circles ('o') on these two curves. The epoch interval at any circle is the spacing (in number of samples) between this and the next circle. The solid and dotted curves are obtained by joining the epoch interval values.

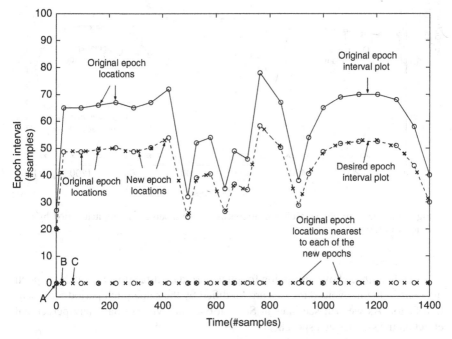

Fig. 6.4 Generation of new sequence of epochs for the modification of pitch period by a factor $\alpha = 0.75$.

Starting from the point A, the new epoch interval value is obtained from the dotted curve, and this value is used to mark the next new epoch B along the x-axis. The epoch interval at this instant on the dotted curve is used to generate the next new epoch C, and so on. The new epoch sequence is marked as '×' along the dotted curve, and also along the x-axis. The nearest original epoch for each of these new epochs is also marked as a sequence of circles ('o') along the x-axis.

Fig. 6.3 illustrates the generation of new epoch sequence when the pitch period is scaled up by $\alpha = 1.5$. The procedure is similar to the one used for the case of Fig. 6.4, except that the new epoch interval values are obtained from the scaled up plot (shown by the dotted curve).

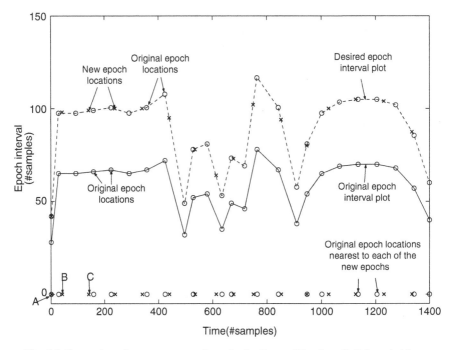

Fig. 6.5 Generation of new sequence of epochs for the modification of pitch period by a factor $\alpha = 1.5$.

Note that in the above discussion for pitch period modification, the random epoch intervals in the nonvoiced regions are modified by the same pitch period modification factor. As we will see later in Section 6.6, this will not have any perceptual effect on the synthesized speech.

6.4 Duration Modification

Generation of new epoch sequence for duration modification is illustrated in Fig. 6.6 for a duration increase by $\beta = 1.5$ times, and in Fig. 6.7 for a duration decrease by $\beta = 0.75$ times. For generating the desired epoch interval plot for duration modification, the original epoch interval plot (solid lines in Figs. 6.6 and 6.7) is resampled according to the desired modification factor.

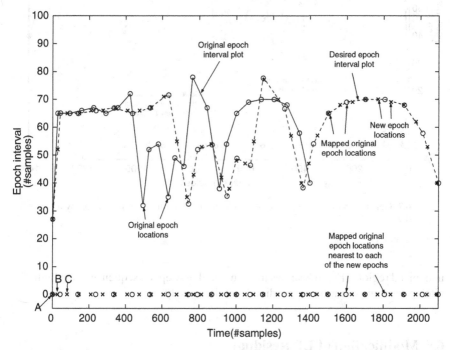

Fig. 6.6 Generation of new sequence of epochs for the modification of duration by a factor $\beta=1.5$.

The desired epoch interval plot is shown by the dotted curve. The modified (new) epoch sequence is generated as follows. Starting with the point A in Fig. 6.6, the epoch interval value is obtained from the dotted curve, and it is used to determine the next epoch instant B. The value of the next epoch interval at B is obtained from the dotted curve, and this value is used to mark the next new epoch C. The new epochs generated by this process are marked as 'x' along the x-axis in Fig. 6.6. The new epochs are also marked ('x') on the desired epoch interval plot along with the mapped original epochs ('o'). Those mapped original epochs nearest to the new epochs are shown along the x-axis by circles ('o'). In a similar manner, the new epochs are generated for the case of decrease of duration, and are shown in Fig. 6.7. Note that in this case also, no distinction is made for epochs in the voiced and

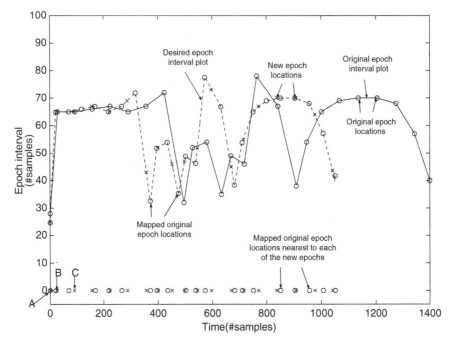

Fig. 6.7 Generation of new sequence of epochs for the modification of duration by a factor $\beta = 0.75$.

nonvoiced regions. Fig. 6.8 shows the generated new epoch sequence when both the pitch period and duration are modified simultaneously.

6.5 Modification of LP Residual

After obtaining the modified epoch sequence, the next step is to derive the excitation signal or LP residual. For this, the original epoch (represented by a 'o') closest to the modified epoch ('×') is determined from the sequence of 'o' and '×' along the desired epoch interval curve (dotted curves in each of the Figs. 6.4 to 6.8). As mentioned earlier, with each original epoch, i.e., the circles ('o') in the plots, there is an associated LP residual sequence of length equal to the value of the original epoch interval for that epoch. The residual samples are placed starting from the corresponding new epoch. Since the value of the desired epoch interval (M) is different from the value of the corresponding original epoch interval (N), there is a need either to delete some residual samples or append some new residual samples to fill the new epoch interval.

Increasing or decreasing the number of LP residual samples for pitch period modification can be done in two ways. In the first method, all the residual samples are used to resample them to the required number of new samples. While there is no

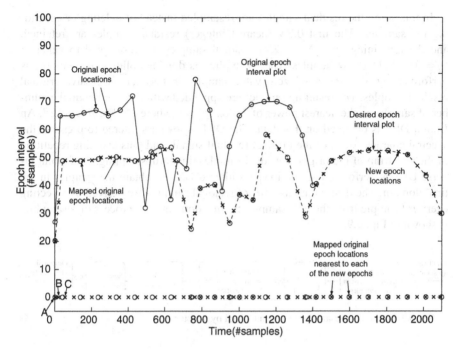

Fig. 6.8 Generation of new sequence of epochs for the modification of pitch period by a factor $\alpha = 0.75$ and duration by a factor $\beta = 1.5$.

distortion perceived in the synthetic speech, the residual samples are expanded or compressed even in the crucial region around the instant of glottal closure. In the second method, a small percentage of the residual samples within a pitch period are retained (i.e., they are not modified), and the rest of the samples are expanded and compressed depending on the pitch period modification factor. The residual samples to be retained are around the instant of glottal closure, as these samples determine the strength and quality of excitation. Thus, by retaining these samples, we will be preserving the naturalness of the original voice.

The percentage of samples to be retained around the instant of glottal closure may not be critical, but if we use a small number (say less than 10% of pitch period) of samples, then we may miss some crucial information in some pitch periods, especially when the period is small. On the other hand, if we consider large number (say about 30%) of samples, then we may include the complete glottal closure region, which will not change in proportion when the pitch period is modified.

The effect of retaining the percentage of the residual samples is examined by subjective evaluation. Three cases have been considered namely, 0%, 20% and 33% of the residual samples to be retained around the instant of glottal closure. No significant difference was perceived in the quality of the synthetic speech. In fact listening tests gave nearly the same level of confidence in all the cases. We have chosen to retain 20% of the residual samples in this study.

In this study the residual samples are resampled instead of deleting or appending the samples. The first $0.2N$ (nearest integer) residual samples are retained, and the remaining $(p = N - 0.2N)$ residual samples are resampled to generate $(q = M - 0.2N)$ new samples. The resampling is done as follows: Resampling is performed by inserting $q - 1$ zero value samples in between successive original residual samples. The resulting samples are appended with zeros to obtain the number of samples to the nearest power of 2, i.e., $2^m = l$, where $2^{m-1} < p*q < 2^m$. An l-point DFT is obtained on this data. The DFT is lowpass filtered to preserve the spectral characteristics of the original residual signal, and thus avoiding repetition of the spectrum of the original residual samples due to upsampling. An l-point inverse DFT is performed on the lowpass filtered DFT to obtain the samples in the time domain. The desired number (q) of residual samples are derived by selecting every p^{th} sample from the new samples in time domain. The process of resampling is shown in Fig. 6.9.

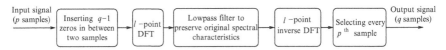

Fig. 6.9 Method for resampling the input signal by a factor q/p.

6.6 Generating the Synthetic Signal

The modified LP residual signal is used as an excitation signal for the time varying all-pole filter. The filter coefficients are updated for every P samples, where P is the frame shift used for performing the LP analysis. In these studies a frame shift of 5 ms and a frame size of 20 ms are used for LP analysis. Thus the P samples correspond to 5 ms when the prosody modification does not involve any duration modification. On the other hand, if there is a duration modification by a scale factor β, then the filter coefficients (LPCs) are updated for every P samples corresponding to 5β ms.

Since the LP residual is used for incorporating the desired prosody modification, there is no significant distortion due to resampling the residual samples both in the voiced and in the nonvoiced regions. This is because there is less correlation among samples in the LP residual compared to the correlation among the signal samples.

Figs. 6.10 and 6.11 show the speech waveforms and narrowband spectrograms for pitch period modification and duration modification, respectively. It can be noted that there are no significant discontinuities in the synthesized speech waveforms or in their spectrograms. Most of the features (such as pitch changes and formant transitions) seem to have been preserved well. This is also verified by perceptual studies discussed in Section 6.7.

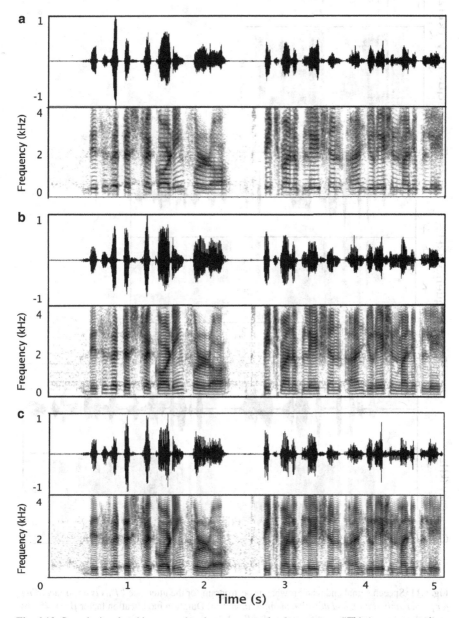

Fig. 6.10 Speech signal and its narrowband spectrogram for the utterance *"This is a test recording for pitch manipulation and duration manipulation"*. (a) Pitch period modification factor $\alpha = 0.66$, (b) original, and (c) pitch period modification factor $\alpha = 1.33$.

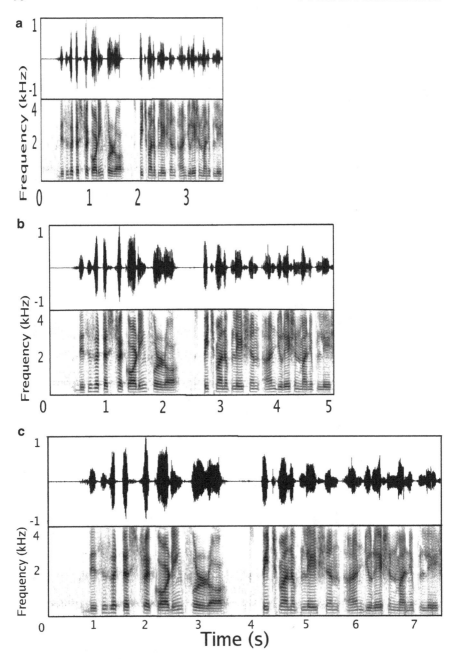

Fig. 6.11 Speech signal and its narrowband spectrogram for the utterance *"This is a test recording for pitch manipulation and duration manipulation"*. (a) Duration modification factor $\beta = 0.75$, (b) original, and (c) duration modification factor $\beta = 1.5$.

6.7 Comparison of the Performance of Proposed Method with LP-PSOLA Method

The performance of the proposed epoch-based prosody modification method is compared with that of the LP-PSOLA method, since the latter method also performs pitch period and duration modification by manipulating the LP residual. The LP-PSOLA method performs pitch and time-scale modifications using pitch markers as anchor points. Here, the instants of significant excitation (epochs) derived by the group-delay analysis are used as the pitch markers for performing the pitch period and duration modification. The process of residual manipulation used in the LP-PSOLA method is briefly described below.

The LP residual signal is divided into overlapping short-time segments by multiplying the LP residual signal by a sequence of pitch-synchronous analysis windows. A Hann window is used in the analysis, and the window is centered around the pitch marker (the instant of glottal closure) for each pitch period. The size of the window is about 2 pitch periods, where the pitch period is estimated as the average of two intervals of the pitch markers on either side of the current pitch marker. New pitch markers are generated from the pitch markers of the given speech signal according to the desired pitch period modification factor. In the case of pitch period modification, the residual signal is manipulated by positioning the windowed segments centered around the new pitch markers, and then adding the overlapped regions. For maintaining the length of the speech signal same before and after the pitch period modification, some of the windowed residual segments are dropped in the case of increasing pitch period, and some of the segments are replicated in the case of decreasing pitch period. Fig. 6.12 shows the manipulation of the LP residual for pitch period modification using the LP-PSOLA method. For duration modification the interval between the pitch markers is not changed, and new time-scaled pitch markers are derived according to the desired modification factor. Finally, the duration modification is realized by deleting or replicating some of the windowed residual segments.

Perceptual evaluation was carried out by conducting subjective tests with 25 research scholars in the age group of 25-35 years. The subjects have sufficient speech knowledge for proper assessment of the speech signals, as all of them have taken a full semester course on speech technology. Four sentences were randomly chosen from the TIMIT database to perform the test [155]. The sentences were spoken by two male and two female speakers. For each sentence the pitch period was modified by factors 2, 1.33, 0.66, 0.5, and 0.4, which corresponds to pitch frequency modification factors 0.5, 0.75, 1.5, 2 and 2.5, respectively. After modification using the proposed method and the LP-PSOLA method, the file names were coded to avoid bias towards a specific method. Each of the subjects were given a pilot test about perception of speech signals for different pitch period modification factors. Once they were comfortable with judging, they were allowed to take the tests. The tests were conducted in the laboratory environment by playing the speech signals through headphones. In the test, the subjects were asked to judge the distortion and

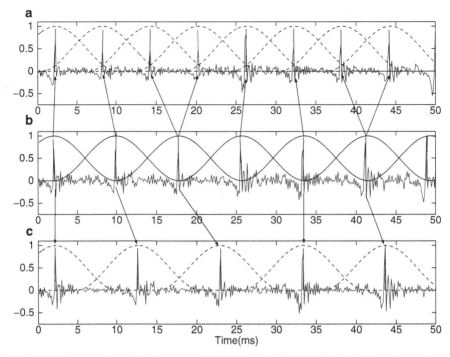

Fig. 6.12 Pitch period modification using LP-PSOLA method. (a) Modified residual signal using LP-PSOLA method for pitch period modification factor $\alpha = 0.75$. (b) LP residual for a segment of voiced speech. (c) Modified residual signal using LP-PSOLA method for pitch period modification factor $\alpha = 1.33$.

quality of the speech for various modification factors. Subjects were asked to assess the quality and distortion on a 5-point scale for each of the sentences obtained by both the methods. The 5-point scale for representing the quality of speech and the distortion level is given in Table 6.2 [156]. Altogether each subject had to judge 40 sentences.

Table 6.2 Ranking used for judging the quality and distortion of the speech signal modified by different modification factors.

Rating	Speech quality	Level of distortion
1.	Unsatisfactory	Very annoying and objectionable
2.	Poor	Annoying but not objectionable
3.	Fair	Perceptible and slightly annoying
4.	Good	Just perceptible but not annoying
5.	Excellent	Imperceptible

The Mean Opinion Scores (MOSs) for each of the pitch period modification factors are given in Table 6.3. The significance of the differences in the pairs of the

MOSs is tested using hypothesis testing [157]. The level of confidence for the observed differences in the sample means was obtained in each case using the sample variances and values of Student-t distribution. The level of confidence is high ($> 97.5\%$) in all cases. This indicates that the differences in the pairs of the MOSs in each case is significant. Hence the epoch-based (proposed) method is prefered over the LP-PSOLA method by all the subjects.

Table 6.3 Mean opinion scores and confidence values for different pitch period modification factors.

Pitch period modification factor (α)	Mean opinion score (MOS)		Level of confidence in % for the significance of difference in MOSs
	LP-PSOLA method	Epoch-based method	
2	3.38	3.94	> 99.5
1.33	4.22	4.48	> 99.5
0.66	4.39	4.57	> 97.5
0.5	3.93	4.42	> 99.5
0.4	3.79	4.19	> 99.5

The scores also indicate that the performance is low for the pitch period modification factors 2 and 0.4. For small modifications (0.66 and 1.33) both the methods seem to perform equally well. For the pitch period modification factor of 2, the performance of the LP-PSOLA method is slightly inferior to that of the epoch-based modification method. In the LP-PSOLA method increase in the pitch period upto a factor of 2 can be achieved using a segment of window length of 2 pitch periods. For further increase in the pitch period, the window size should include more than two pitch periods. For instance, using a window size of four pitch periods, the pitch period can be increased upto 4 times. However, with large window sizes secondary excitations are introduced. For the modification at the lower end (less than 0.5), the LP-PSOLA introduces some audible distortion, whereas the epoch-based method gives lesser distortion. The LP-PSOLA method produces phase mismatches and audible distortion due to overlap and adding of the windowed residual segments, for both increasing and decreasing pitch period cases.

To visualize these distortions, the residual manipulation using the epoch-based method and the LP-PSOLA are illustrated for different pitch period modification factors. Fig. 6.13 shows the LP residual for a segment of voiced speech, and the residuals modified for pitch period modification factors of 2 and 0.5 using the LP-PSOLA method and the proposed epoch-based method. From the figure we can see that, after modification, the general characteristics of the LP residual are preserved better in the epoch-based approach, compared to the results from the LP-PSOLA method. These changes in the residual are reflected in the results of the listening tests for the pitch period modification factors of 2, 0.5 and 0.4.

For evaluating the proposed method for duration modification, a similar approach was followed as in the case of pitch period modification. The mean opinion scores

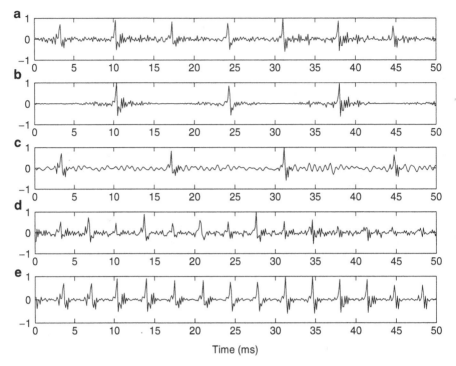

Fig. 6.13 (a) LP residual for a segment of voiced speech. (b) Modified LP residual signal using LP-PSOLA method for pitch period modification factor $\alpha = 2$. (c) Modified LP residual signal using epoch-based method for $\alpha = 2$. (d) Modified LP residual signal using LP-PSOLA method for $\alpha = 0.5$. (e) Modified LP residual signal using epoch-based method for $\alpha = 0.5$.

for different duration modification factors are given in Table 6.4. The significance of differences in the pairs of MOSs is tested using hypothesis testing in this case also. In this case the differences in the pairs of MOSs is not significant for most of the duration modification factors, as the percentage confidence was only about 90%. The scores show that both the methods seem to perform equally well for duration modification.

Table 6.4 Mean opinion scores and confidence values for different duration modification factors.

Duration modification factor (β)	Mean opinion score (MOS) LP-PSOLA method	Epoch-based method	Level of confidence in % for the significance of difference in MOSs
0.5	3.78	3.63	> 90
0.75	4.57	4.63	< 90
1.5	4.65	4.71	< 90
2	4.12	4.37	99.5
2.5	3.96	4.07	< 90

The performance of the proposed prosody modification method is also compared with the performance of Time Domain Pitch Synchronous Overlap and Add (TD-PSOLA) method, since TD-PSOLA method is more popular and widely used in speech synthesis systems. The details of TD-PSOLA method were discussed in Chapter 2. Perceptual evaluation is carried out in a way similar to the comparison performed with LP-PSOLA method. The MOSs and level of confidence for the differences in the pairs of MOSs for each of the pitch period and duration modification factors are given in Tables 6.5 and 6.6, respectively. The MOSs and confidence values indicate that performance of the epoch-based method is better than TD-PSOLA method. In TD-PSOLA method, both for pitch period modification and duration modification, we can observe the spectral and phase distortion for larger modification factors.

Table 6.5 Mean opinion scores and confidence values for different pitch period modification factors.

Pitch period modification factor (α)	Mean opinion score (MOS)		Level of confidence in % for the significance of difference in MOSs
	TD-PSOLA method	Epoch-based method	
2	3.23	3.94	> 99.5
1.33	4.27	4.48	> 99.5
0.66	4.21	4.57	> 99.5
0.5	3.71	4.42	> 99.5
0.4	3.62	4.19	> 99.5

Table 6.6 Mean opinion scores and confidence values for different duration modification factors.

Duration modification factor (β)	Mean opinion score (MOS)		Level of confidence in % for the significance of difference in MOSs
	TD-PSOLA method	Epoch-based method	
0.5	3.66	3.63	< 90
0.75	4.46	4.63	> 90
1.5	4.59	4.71	> 90
2	4.16	4.37	99.5
2.5	3.87	4.07	> 95

6.8 Summary

A flexible method for manipulating the prosody (pitch and duration) parameters of a speech utterance was proposed [158–160]. The method uses the features of source of excitation of the vocal tract system. The linear prediction residual was used to represent the excitation information. The prosody manipulation was performed by extracting the instants of significant excitation (epochs) from the LP residual, and generating a new epoch sequence according to the desired prosody modification. A modified LP residual was generated using the knowledge of the new epoch sequence. In generating this residual, the perceptually significant portion (20% of the region around the instant of glottal closure) was retained, and the remaining 80% of the residual samples were used to generate the required number of samples in the modified residual. It is interesting to note that the epochs in both the voiced and nonvoiced regions are treated alike, thus avoiding a separate voiced, unvoiced and silence (V/UV/S) decision making. Also since the manipulation was performed on the residual signal, distortions were not perceived. This is because the residual samples are less correlated than the signal samples. This feature also helps in realizing prosody modification by large modification factors. The modification procedure is similar both for pitch period and for duration.

The modified LP residual signal was used as the excitation signal for the time varying all-pole filter to synthesize the speech for the desired prosody. The performance of the proposed prosody modification method was compared with familiar methods, namely, LP-PSOLA and TD-PSOLA. Perceptual evaluation showed that performance of the proposed epoch-based method was superior compared to LP-PSOLA and TD-PSOLA methods. The applications based on prosody modification and some of the issues in using the proposed method are discussed in the next chapter.

Chapter 7
PRACTICAL ASPECTS OF PROSODY MODIFICATION

Abstract This chapter discuss about the practical aspects of prosody modification. A computationally efficient method for determining the instants of significant excitation is proposed. The application of proposed duration and intonation prediction models has been demonstrated using concatenative text to speech synthesis system. A new duration modification method using vowel onset points and instants of significant excitation is proposed. For voice conversion application, methods are developed to modify the formant structure and to impose the desired pitch contour on a given speech signal.

7.1 Introduction

This chapter presents some of the practical issues that arise when the proposed prosody modification method discussed in the previous chapter is used for different applications. One application of prosody modification method lies in enhancing the intelligibility of the degraded speech. The intelligibility of degraded speech can be significantly improved by increasing its duration. This can be achieved by prosody modification. For online applications such as Text-to-Speech (TTS) synthesis and voice conversion, it is necessary to modify the prosody parameters in real time. In the method proposed for prosody modification, most of the computation effort is in determining the instants of significant excitation. A computationally efficient method is proposed in this chapter to determine the instants of significant excitation from speech signal. Since the proposed method modifies the durations of all speech regions uniformly, larger modification factors produce distortion due to unnatural expansion/compression of consonant and transition regions. A new method for duration modification is proposed in this chapter to provide flexibility for modification in different regions. In voice conversion application besides prosody, it is necessary to modify both the excitation source and the vocal tract system parameters according

K.S. Rao, *Predicting Prosody from Text for Text-to-Speech Synthesis*, SpringerBriefs
in Electrical and Computer Engineering, DOI 10.1007/978-1-4614-1338-7_7,
© Springer Science+Business Media New York 2012

to the specification of the target speech. Methods are proposed in this chapter to modify the formant structure, and to impose the desired pitch contour on a given speech signal.

This chapter is organized as follows: A computationally efficient method to determine the instants of significant excitation from speech signal is discussed in Section 7.2. Section 7.3 briefly discuss the application of both prosodic prediction and incorporation methods in the context of text-to-speech synthesis. In Section 7.4, a new duration modification method is proposed, which provides flexibility for modification in different regions. Issues in using the prosody modification method for voice conversion are discussed in Section 7.5.

7.2 A Computationally Efficient Method for Extracting the Instants of Significant Excitation

Determining the instants of significant excitation using group delay based method is a computationally intensive process, since the group delay is computed for every sample shift. Computation of group delay for each frame involves the computation of two discrete Fourier transforms. The computational complexity can be reduced to some extent by computing the group delay only for the samples around the instants of Glottal Closure (GC). This is achieved by first detecting approximate locations of the glottal closure instants. The peaks in the Hilbert envelope of the Linear Prediction (LP) residual indicate the approximate locations of the GC instants [161].

Even though the real and imaginary parts of an analytic signal (related through the Hilbert transform) have positive and negative samples, the Hilbert envelope of the signal is a positive function, giving the envelope of the signal [162]. The properties of Hilbert envelope can be exploited to detect approximate locations of the GC instants. The Hilbert envelope $h_e(n)$ of the LP residual $e(n)$ is defined as follows [161–163]:

$$h_e(n) = \sqrt{e^2(n) + e_h^2(n)}$$

where $e_h(n)$ is the Hilbert transform of $e(n)$, and is given by [161]

$$e_h(n) = IFT[E_h(\omega)]$$

where

$$E_h(\omega) = \begin{cases} -jE(\omega), & 0 \leq \omega < \pi \\ jE(\omega), & -\pi \leq \omega < 0 \end{cases}$$

Here IFT denotes the Inverse Fourier Transform, and $E(\omega)$ is the Fourier transform of $e(n)$. Fig. 7.1 shows a segment of voiced speech, its LP residual, Hilbert transform and the Hilbert envelope. The peaks in the Hilbert envelope indicate epoch locations.

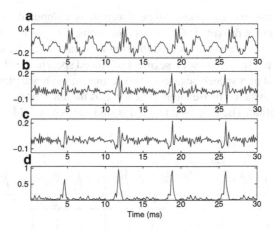

Fig. 7.1   (a) A segment of voiced speech, corresponding (b) LP residual, (c) Hilbert transform of the LP residual, and (d) Hilbert envelope of the LP residual.

The evidence of glottal closure instants is obtained by convolving the Hilbert envelope with a Gabor filter (modulated Gaussian pulse) given by, $g(n) = \frac{1}{\sqrt{2\pi}\sigma}e^{-\frac{n^2}{2\sigma^2}+j\omega n}$. Where σ defines the spatial spread of the Gaussian, ω is the frequency of modulating sinusoid [164], and n is the length of the filter. The Gabor filter used in this study is shown in Fig. 7.2.

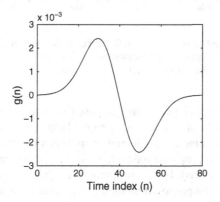

Fig. 7.2   Gabor window for $\sigma = 10$, $\omega = 0.0114$ and $n = 80$.

The plot of the evidence is termed as the *GC Evidence Plot*. In the GC evidence plot, the instants of positive zero–crossings correspond to approximate locations of the instants of significant excitation. To determine the accurate locations of the glottal closure instants, the phase slope function is computed for the residual samples around the approximate GC instant locations. The positive zero–crossings of the phase slope function correspond to accurate locations of the instants of significant excitation. Fig. 7.3 shows a segment of voiced speech, the Hilbert envelope of the LP residual of a speech segment, the GC instant evidence plot, approximate locations of GC instants, phase slope function and the locations of the instants of significant excitation. The proposed method is summarized in Table 7.1.

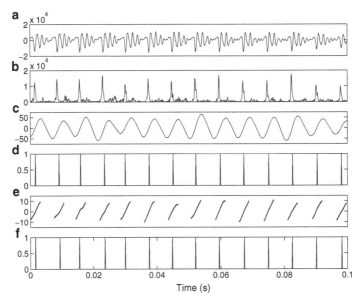

Fig. 7.3 (a) A segment of voiced speech, (b) Hilbert envelope of the LP residual, (c) GC instant evidence plot, (d) approximate locations of GC instants, (e) phase slope function and (f) accurate locations of the instants of significant excitation.

The computational efficiency of the proposed method depends on the number of approximate epoch locations derived from the Hilbert envelope of the LP residual and the number of samples considered around each GC instant. For evaluating the performance of the proposed method, 10 speech utterances, each of duration of 3 seconds are considered. For each utterance the instants of significant excitation are computed by the proposed method using different window sizes (number of samples around the approximate instant location). The epochs determined by the standard group delay method are used as reference [152]. Table 7.2 shows the number of instant locations derived by the proposed method for different window sizes. The total number of instants derived from all utterances using the group delay method is 2477. The total number of approximate instant locations for all the utterances derived by using the Hilbert envelope of the LP residual is 2673. It is observed

Table 7.1 Steps for automatic detection of the instants of significant excitation using Hilbert envelope of LP residual and group delay function.

1. Preemphasize input speech.
2. Compute LP residual with 10th order LP analysis, frame size of 20 ms and shift of 5 ms.
3. Compute Hilbert envelope of the LP residual, enhance the peaks in Hilbert envelope by dividing each sample of the HE with the running mean around that sample
4. Obtain the GC instant evidence plot by convolving the enhanced HE with Gabor filter.
5. Find the positive zero-crossing locations in the GC instant evidence plot, which are hypothesized as the approximate locations of the instants of significant excitation.
6. Compute the group delay for the samples within 2 ms window around the approximate GC instant locations.
7. Derive the phase slope function from group delay values, and smooth it with a Hamming window.
8. Identify positive zero-crossings of the phase slope as the accurate instants of significant excitation.

that when the size of the window is small, the computational efficiency is high but at the same time, some of the epochs will be missing. As the size of the window increases, the computational efficiency decreases, but the number of missing epochs also decreases.

Table 7.2 Number of instants derived using the proposed method for different window sizes.

Window size (ms)	# Instants	% Instants
0.5	1563	63
1.0	2240	90
1.5	2373	96
2.0	2406	97
2.5	2429	98
3.0	2445	99
3.5	2457	99
4.0	2460	99

The deviation in the approximate epoch locations with respect to their reference locations are computed. The results of the analysis are given in Table 7.3. The entries in the Table 7.3 indicate the number of approximate instants and their deviation in terms of number of samples with respect to reference instants. On the whole the average deviation per instant is found to be 2.3 samples (0.29 ms).

Table 7.3 Number of approximate instants derived from Hilbert envelope for different deviations with respect to the locations of reference instants.

Deviation (# samples)	# Instants	% Instants
0	535	22
1	622	25
2	406	16
3	453	18
4	224	9
5	92	4

7.3 Text-to-speech synthesis

For Text-to-Speech (TTS) synthesis, prosody models (duration and intonation models) provide the specific duration and intonation information associated with the sequence of sound units present in the given text. In this study we employ waveform concatenation for synthesizing the speech [165, 166]. The basis of concatenative synthesis is to join short segments of speech, usually taken from a pre-recorded database, and then impose the associated prosody by appropriate signal processing methods. We considered syllable as a basic unit for synthesis. The database used for concatenative synthesis consists of syllables, which are extracted from a carrier words spoken in isolation. We have selected nonsense words as carrier words, because they offer minimum bias towards any of the factors that affect on the basic characteristics of the syllable. These syllables are considered as neutral syllables. Using some guidelines the carrier words can be formed for different categories of syllables. Some of the guidelines are prepared while developing the TTS system for Hindi at IIT Madras [166].

In text-to-speech synthesis, we need to synthesize the speech for the given text. Firstly, the given text is analyzed using a text analysis module, and then the positional, contextual and phonological features (linguistic and production constraints) for each of the syllables present in the given text are derived. These features are presented to the duration and intonation models (see Section 4.2), which will generate the appropriate duration and intonation information corresponding to the syllables.

At synthesis stage, firstly, the waveforms of the pre-recorded (neutral) syllables are concatenated according to the sequence present in the text. The derived duration and intonation knowledge, corresponding to the sequence of syllables is incorporated in the sequence of concatenated syllables using prosody modification methods [158]. The prosody parameters (duration and pitch) are incorporated by manipulating the instants of significant excitation of the vocal tract system during the production of speech [4]. Instants of significant excitation are computed from the Linear Prediction (LP) residual of the speech signals by using the average group-delay of minimum phase signals [151, 152].

The quality (intelligibility and naturalness) of synthesized speech is evaluated by perceptual studies. Perceptual evaluation is performed by conducting subjective tests with 25 research scholars in the age group of 25-35. The subjects have sufficient speech knowledge for proper assessment of the speech signals. Five sentences are synthesized from text for each of the languages Hindi, Telugu, Tamil and Kannada. Each of the subjects were given a pilot test about perception of speech signals with respect to intelligibility and naturalness. Once they are comfortable with judging, they were asked to take the tests. The tests were conducted in a laboratory environment by playing the speech signals through headphones. In the test, the subjects were asked to judge the intelligibility and naturalness of the speech. Subjects have to assess the quality on a 5-point scale for each of the sentences. The 5-point scale for representing the quality of speech is given in Table 7.4 [156].

Table 7.4 Ranking used for judging the quality of the speech signal.

Rating	Speech quality
1.	Unsatisfactory
2.	Poor
3.	Fair
4.	Good
5.	Excellent

The mean opinion scores (MOS) for assessing the intelligibility and naturalness of the synthesized speech in each of the languages Hindi, Telugu, Tamil and Kannada are given in columns 2 and 3 of Table 7.5. The scores indicate that the intelligibility of the synthesized speech is fairly acceptable for all the languages, whereas the naturalness seems to be poor. Naturalness is mainly attributed to individual perception. Naturalness can be improved to some extent by incorporating the coarticulation and stress information along with duration and intonation.

For analyzing the accuracy of the prediction of prosody models, we also conducted the listening tests for assessing the intelligibility and naturalness on the speech without incorporating the prosody. In this case, speech samples are derived by concatenating the neutral syllables without incorporating the prosody. The mean opinion scores of these listening tests are given in columns 4 and 5 of Table 7.5. The MOS of the quality of the speech without incorporating the prosody have been observed to be low compared to the speech synthesized by incorporating the prosody. The significance of the differences in the pairs of the mean opinion scores for intelligibility and naturalness is tested using hypothesis testing [157]. The level of confidence for the observed differences in the sample means was obtained in each case using the sample variances and values of Student-t distribution. The level of confidence is high ($> 99.5\%$) in all cases (shown in columns 6 and 7 of Table 7.5). This indicates that the differences in the pairs of MOS in each case is significant.

Table 7.5 Mean opinion scores for the quality of synthesized speech in the languages Hindi, Telugu, Tamil and Kannada. (Intl: Intelligibility, Nat: Naturalness)

Language	Mean opinion score (MOS)				Level of) confidence (%)	
	TTS with prosody		TTS without prosody			
	Intl	Nat	Intl	Nat	Intl	Nat
Hindi	3.87	2.93	2.91	2.03	> 99.5	> 99.5
Telugu	4.15	3.24	3.27	2.32	> 99.5	> 99.5
Tamil	4.19	3.18	3.13	2.29	> 99.5	> 99.5
Kannada	3.95	3.12	2.97	2.23	> 99.5	> 99.5

7.4 Duration Modification using Vowel Onset Points and Glottal Closure Instants

Most of the time scale modification methods vary the duration of speech uniformly over all regions. But it is observed that the transition regions between a consonant and the following vowel, and the consonant regions in the speech signal do not vary appreciably with speaking rate [106, 167]. Therefore a method is proposed to modify duration of an utterance without changing the durations of transition and consonant regions. Vowel Onset Points (VOPs) are used to identify the transition and consonant regions. Vowel onset point corresponds to the instant at which the onset of vowel takes place, i.e., the transition from the consonant to vowel in most cases. VOPs are determined using the Hilbert envelope of the linear prediction residual of speech signal [168].

For studying the characteristics of the naturally spoken utterances at different speeds (fast, normal and slow), various sentences uttered by different subjects at different speaking rates are manually analyzed. The analysis was performed as follows: Five Hindi sentences were chosen for this study. These sentences were uttered by ten cooperative subjects in laboratory environment. Each sentence was uttered three times by each subject: (1) At normal speaking rate, (2) at slow rate and (3) at fast rate. Altogether 150 sentences were recorded using 5 different sentences uttered in three different speaking rates by 10 subjects. The sentences were analyzed in the transition regions, vowel regions, consonant regions and pauses. It is observed that in most of the cases, vowels and pauses are affected by speaking rate variations, whereas the consonant and transition regions remain mostly unaltered. Therefore we propose a method for duration modification using the vowel onset points and glottal closure instants, which allows modification of speech duration excluding the transition and consonant regions.

7.4.1 Detection of VOPs

Speech signal is sampled at 8 kHz and preemphasized before performing LP analysis. The LP residual is computed using 10th order LP analysis, with a frame size of 20 ms and a frame shift of 5 ms. The Hilbert envelope of the LP residual is then computed. The VOP evidence is obtained from the Hilbert envelope of the LP residual by convolving it with a Gabor filter. A Gabor filter with parameter values, spatial spread of the Gaussian $\sigma = 100$, the frequency of modulating sinusoid $\omega = 0.0114$, and a filter length $n = 800$, is considered [164]. In the VOP evidence plot the peaks are located using a peak picking algorithm. Spurious peaks are eliminated using the characteristics of the shape of the VOP evidence plot, namely, between two true VOPs, there exists a negative region of sufficient strength due to vowel region. For each peak, a check for the presence of such a negative region with respect to next peak is made, to eliminate the spurious peaks [168].

The above procedure is illustrated for the Hindi sentence "*antarAshtriyA bassevA pichale mahIne shuru huyi thI*". In this sentence there are 16 VOP events, as marked (manually) in Fig. 7.4(a). The Hilbert envelope and the VOP evidence plots are shown in Figs. 7.4(b) and (c), respectively. The output of the peak picking algorithm is given in Fig. 7.4(d). The hypothesized VOPs after eliminating the spurious ones are shown in Fig. 7.4(e). The procedure for detecting the VOPs in speech signals is summarized in Table 7.6.

Table 7.6 Steps for detection of the VOP events.

1. Preemphasize input speech.
2. Compute LP residual with 10th order LP analysis, with a frame size of 20 ms and shift of 5 ms.
3. Compute Hilbert envelope of the LP residual
4. Obtain the VOP event evidence plot from Hilbert envelope by convolving it with the Gabor filter.
5. Identify the peaks in the VOP event evidence plot using peak picking algorithm.
6. For each peak, if there is no negative region with reference to next peak, then eliminate such a peak as spurious.
7. Eliminate peaks which are at a distance less than 50 ms with respect to their next peak.
8. Hypothesize remaining peaks as the VOP events.

Vowel onset point can be interpreted as the junction point between consonant and vowel of a CV unit. The region to the left of the VOP is considered as the consonant region, and to the right of the VOP as the vowel region. In the vowel portion, a small region following the VOP is treated as transition region [169]. After determining the vowel onset point, 30 ms to the left of VOP is marked as consonant region, and 30

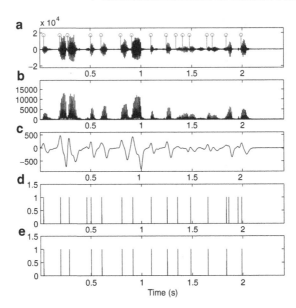

Fig. 7.4 Hindi sentence *"antarAshtriyA bassevA pichale mahlne shuru huyi thi"* (a) Speech waveform with manual marked VOP events, (b) Hilbert envelope of LP residual, (c) VOP evidences, (d) output of peak picking algorithm, and (e) hypothesized VOP events after eliminating some spurious peaks.

ms to the right of VOP is marked as transition region. In the proposed modification method, the durations of these regions are not modified.

7.4.2 Modification of duration

Using the prosody modification method, the LP residual is modified according to the required modification factor (β). Vowel onset points are determined for a given speech utterance using the procedure mentioned in Section 7.4.1. The new VOPs corresponding to the modified LP residual are generated by scaling the VOP locations by the desired modification factor. For preserving the consonant and transition regions in speech signal during modification, the scaled transition and consonant regions in the modified residual are replaced by the corresponding regions in the original LP residual signal. The filter coefficients (LP coefficients) are updated depending on the length of the modified LP residual. Speech for the desired duration modification can be synthesized by exciting the all-pole filter using the modified LP residual.

7.4.3 Perceptual evaluation

The performance of the proposed duration modification method is compared with the performance of the conventional (uniform) duration modification method using perceptual evaluation. Here the conventional duration modification is implemented by the prosody modification method discussed in Chapter 6. The details of subjective tests to carry out perceptual evaluation were given in Chapter 6. Two sentences (spoken by male and female speakers) were chosen for analysis. Speech signals were derived for the modification factors from 0.25 to 3 in the steps of 0.25. The Mean Opinion Scores (MOSs) for each of the modification factors are obtained. The mean opinion scores for different ranges in modification are shown in Table 7.7. The level of confidence is computed for the difference of each pair of MOSs. Results of perceptual evaluation show that at lower modification rates both uniform and proposed duration modification methods produce a good quality speech. When the modification factors reach around 0.5 (compression) and 2 (expansion), the conventional method produce slight distortion, whereas the proposed method provides intelligible and better quality speech. Speech signals used for subjective tests are available for listening at $http://speech.cs.iitm.ernet.in/Main/result/prosody.html$.

Table 7.7 Mean opinion scores and confidence values for different intervals of modification factors.

Duration modification interval	Mean opinion score		Level of confidence in % for the significance of difference in MOSs
	Conventional method	Proposed method	
< 0.75	2.78	3.52	> 99.5
0.75 to 1.5	4.20	4.36	> 90
> 1.5	3.36	4.13	> 99.5

7.5 Voice Conversion

In voice conversion, the intonation and durational patterns of the target speaker needs to be incorporated. That is, the pitch information need to be modified as per the required pitch contour. The formant structure also needs to be modified along with the pitch according to the specification of the target speaker. This section discusses about modification of the formant structure in accordance with the pitch variations, and presents a method to modify pitch according to the desired pitch contour.

7.5.1 Modification of formant structure

Speaker characteristics lie at the linguistic, suprasegmental and segmental levels [80]. The speaker characteristics at the linguistic and suprasegmental levels are difficult to derive from the data. The speaker characteristics at the segmental level can be attributed to speech production mechanism, and they are reflected in the source and system characteristics [74].

Pitch and formant frequencies represent the unique characteristics for a speaker [74]. For modifying the speaker characteristics, both pitch and formant frequencies need to be altered accordingly. So far in the pitch modification, only the pitch period of the signal is altered by keeping the LP Coefficients (LPCs), unchanged. Therefore the output speech is unnatural due to lack of correspondence between the source and the vocal tract system. To provide naturalness, one should alter the vocal tract shape according to the pitch periodicity. For carrying out this, the interrelation between the source and the vocal tract should be known.

Television (TV) broadcast news data is used for analyzing the relationship between pitch (source) and formant frequencies (vocal tract characteristics) [170]. For the analysis, speech data of three female and two male speakers is considered. The steady vowel regions with identical positional and contextual constraints are derived from the database. Formant frequencies are computed by using the group delay function method [171]. All the group delay functions of a particular vowel are plotted on the same figure, one over the other, as shown in Fig. 7.5. In the figure, the overlapping regions correspond to the formant frequencies of the vowel. The average pitch in the vowel region is also computed. The average pitch and formant frequencies for all vowels and speakers are computed. Table 7.8 shows the dependence of formant frequencies on pitch. In the analysis the first three formants are considered.

Table 7.8 Dependence of formant frequencies on pitch.

Vowel	Male (M)				Female (F)				Ratio (F/M)			
	Pitch	Formant frequencies			Pitch	Formant frequencies			Pitch	f1	f2	f3
		f1	f2	f3		f1	f2	f3				
a	115	652	1410	2365	245	845	1630	2732	2.13	1.30	1.16	1.16
e	132	470	1895	2635	263	605	2165	2915	1.99	1.29	1.14	1.11
i	127	378	2086	2757	239	465	2476	3115	1.88	1.23	1.19	1.13
o	122	562	1315	2515	255	678	1438	2835	2.09	1.21	1.09	1.13
u	118	365	1225	2470	228	475	1357	2992	1.93	1.30	1.11	1.21

The LP coefficients and the LP residual are computed using LP analysis. The LP residual is modified as per the required pitch period modification factor using the method discussed in Chapter 6. Along with the pitch period modification, the shape of the vocal tract is also need to be modified accordingly. The shape of the vocal tract is characterized by the resonances (formant frequencies) of the vocal tract system. The resonances and their bandwidths are related to the angle and magnitude

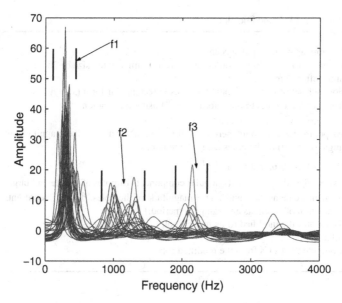

Fig. 7.5 Group delay functions for the vowel /u/.

of the corresponding poles in the z-plane. The formant frequencies can be changed by shifting the poles of a system transfer function in the z-plane [80]. As per the required modification in formant frequencies, the angle and magnitude of the poles are modified. The LPCs are recomputed from these new poles. The basic procedure for the modification of LPCs is given in Table 7.9. After modifying all the LPCs, the system with the new LPCs is excited using the modified residual. The quality of the synthesized speech is close to that of natural speech for the required pitch modification. The original and the modified (pitch and LPCs) speech signals are available for listening at $http://speech.cs.iitm.ernet.in/Main/result/prosody.html$.

7.5.2 Modification of pitch according to pitch contour

So far in the pitch period modification, the pitch contour of the speech utterance is shifted by a constant scale factor. But in some applications like text-to-speech synthesis and voice conversion, the existing pitch contour needs to be modified as per the desired (target) pitch contour. In this study, modification of pitch according to the required pitch contour is also termed as imposing the desired pitch contour on a given speech signal.

For imposing the desired (target) pitch contour on the existing pitch contour, the durations of voiced and nonvoiced regions in both the contours should be equal.

Table 7.9 Steps for LPCs modification.

1 Preemphasize the input speech.
2 Compute LPCs with 10th order LP analysis, with a frame size of 20 ms and shift of 5 ms.
3 For each set of LPCs compute the roots in rectangular form $(x_i \pm jy_i)$.
4 Transform the roots to polar form $(re^{\pm j\theta})$ using the relations
$r = \sqrt{x^2 + y^2}$ and $\theta = \arctan(y/x)$.
5 As per the required pitch period modification factor modify the magnitude and angle (r and θ) of the poles using the relations
$\theta_i' = \alpha_i \theta_i = \frac{f_i'}{f_i} \theta_i$ and $r = e^{-\pi \beta_i T}$.
(where θ_i and θ_i' represents angular components of poles of source and target formant frequencies f_i and f_i', r = magnitude of the poles, β_i and T represents bandwidth of formants and sampling period.)
6 Transform the modified roots into complex conjugate form using the relation $(r\cos\theta + jr\sin\theta)$.
7 Compute the LPCs from the modified roots.

Therefore the voiced and nonvoiced regions of the desired epoch interval plot (desired pitch contour) are resampled according to the requirement. Then, the voiced and nonvoiced regions of the given epoch interval plot are replaced by the resampled voiced and nonvoiced regions of the desired epoch interval plot. Fig. 7.6 shows the modification of the given epoch interval plot according to the desired epoch interval plot.

After modifying the LP residual according to the modified epoch interval plot (imposing the desired pitch contour), the LPCs (filter coefficients of an all-pole filter) of the original signal are excited with this modified residual. The output of the forward filter is the required synthesized speech for the desired pitch contour.

7.6 Summary

For real-time modification of prosody, we need to determine the instants of significant excitation in an efficient manner. A computationally efficient method was proposed for determining the instants of significant excitation from speech, using the Hilbert envelope and group delay function of the LP residual [172, 173]. The method first computes approximate locations of the glottal closure instants using the Hilbert envelope of the LP residual. Then the group delay analysis is performed only for the samples around the glottal closure instants. The significance of the desired prosody for high quality text-to-speech synthesis is demonstrated by using the proposed prosody prediction models and prosody modification methods [129, 144, 174]. Most of the time scale modification methods vary the duration of speech uniformly over all segments of speech. This gives distortion for very large or small

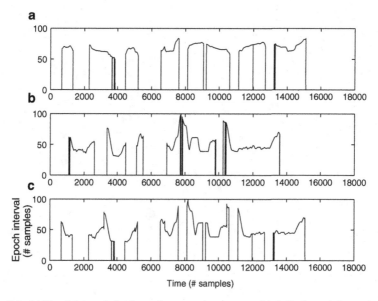

Fig. 7.6 (a) Epoch interval plot for a given speech utterance, (b) desired epoch interval plot and (c) desired epoch interval plot is superimposed on the given epoch interval plot.

modification factors. A new duration modification method was proposed using vowel onset points and glottal closure instants, which provide flexibility for modification in different regions [175]. The formants are related to pitch, and hence when pitch is modified even formants have to be changed accordingly. This is true especially in application such as voice conversion. Methods were developed for incorporating the required formant structure and pitch contour [176, 177]. The formant structure can be modified by shifting the poles of a system transfer function in the z-plane.

Chapter 8
SUMMARY AND CONCLUSIONS

Abstract This chapter summarizes the research work presented in this book, highlights the contributions of the work and discusses the scope for future work.

8.1 Summary of the Work

This book addressed some issues on acquisition and incorporation of prosody knowledge useful for developing speech systems. Acquisition of prosody knowledge involves devising a model that can capture the implicit duration and intonation knowledge present in the speech signal. As a first step towards modeling prosody, a labeled speech corpora of a broadcast news data for Telugu language is analyzed manually. Manual analysis or any rule-based method does not provide the interaction among the linguistic features at various levels. Deriving the rules by analyzing large databases is a tedious process. But, after analyzing the labeled data manually, one can infer different factors that affect the prosody parameters of syllables.

Nonlinear models are known to capture the implicit relation between the input and output. In this book, neural networks and support vector machines are explored for capturing the implicit duration and intonation knowledge present in speech signal. The basic hypothesis is that the values of duration and pitch of a sound unit in speech is related to the linguistic constraints on the unit. The linguistic constraints of a unit are expressed in terms of positional, contextual and phonological information of the unit. For capturing the underlying model of duration and intonation, labeled broadcast news data from the three languages, Hindi, Telugu and Tamil is used as speech corpora. The performance of the models is evaluated by computing the average prediction error (μ), standard deviation (σ) and correlation coefficient (γ). The accuracy of prediction is also analyzed in terms of percentage of syllables predicted within different deviations from their actual values. Prediction performance is improved by postprocessing the predicted values using piecewise

K.S. Rao, *Predicting Prosody from Text for Text-to-Speech Synthesis*, SpringerBriefs
in Electrical and Computer Engineering, DOI 10.1007/978-1-4614-1338-7_8,
© Springer Science+Business Media New York 2012

linear transformation, and then imposing the constraints of the prosody parameters. Preprocessing the input into different stages, results in further improvement in the performance of the models. The performance of the proposed models is improved by exploiting the dependency of the duration and intonation on each other. The performance of these models is compared with the performance of more familiar CART (Classification And Regression Tree) models. The developed prosody models also capture the speaker-specific and language-specific information. This feature may be used in speaker and language identification tasks.

To incorporate the acquired prosody knowledge, a new method of prosody modification of speech is proposed. The method involves deriving a modified excitation signal such as linear prediction (LP) residual. The modification of the excitation signal is based on the knowledge of the instants of significant excitation of the vocal tract system during production of speech. The instants of significant excitation are computed from the linear prediction residual of speech signals using the property of average group delay of minimum phase signals. The group delay based method for extraction of the instants of significant excitation is computationally intensive, since it requires the computation of the group delay for every sample shift. A computationally efficient method is proposed to determine the instants of significant excitation from speech, using the Hilbert envelope and group delay function of the LP residual. The proposed method first computes approximate locations of the glottal closure instants using the Hilbert envelope of the LP residual, and then the group delay analysis is performed only for the samples around the approximate glottal closure instants. Proposed prosody models and prosody modification methods have been successfully applied in syllable based concatenative text-to-speech syntheis for improving the quality of synthesized speech. In general, most of the time scale modification methods vary the duration of speech uniformly over all segments of speech. But it can be observed that, for modification factors greater than 2 or less than 0.5, one notices some distortion in the consonant regions, and in the transition region between a consonant and the following vowel. To circumvent this problem a method is proposed using vowel onset points and glottal closure instants, to provide flexibility for modification in different regions. The proposed prosody modification method can be used for various applications. For instance, in voice conversion we need to modify the intonation and durational patterns of a speaker. The formant structure also needs to be modified along with the pitch according to specification of the target speech. Methods are developed for incorporating the required formant structure and pitch contour.

8.2 Contributions of the Work

The book presents an approach to capture the implicit duration and intonation knowledge using models of neural networks and support vector machines. The results were demonstrated using the labeled database for speech in three Indian languages, namely, Hindi, Telugu and Tamil.

The prosody models were shown to possess speaker and language characteristics as well, besides the information about the message in speech. Thus the models can be explored for speaker and language identification studies.

The major application of prosody models is in text-to-speech synthesis and voice conversion. For this, the prosody knowledge captured by the models need to be incorporated in speech signals. A new method for prosody modification is proposed in this thesis. The method is based on exploiting the nature of excitation of speech production mechanism. In particular, the sequence of the instants of significant excitation (epochs) are derived, and then modified to incorporate the desired prosody. The modified epoch sequence is used to generate an excitation signal to synthesize speech. The synthesized speech sounds natural even for large values of the prosody modification factors. An efficient method for prosody modification is developed for online speech synthesis.

8.3 Scope for Future Work

Duration analysis was performed on broadcast news data in a language Telugu. The analysis was performed systematically by grouping the syllables using size of the word and position of the word in the utterance. This analysis could be extended to other languages, since common and distinctive features across the languages may be useful for several applications.

In this work, linguistic and production constraints are represented with positional, contextual and phonological information. Along with these, information related to accent, prominence, stress and semantics can also to be added to the feature vector. Weighing the constituents of the input feature vector based on linguistic and phonetic importance may further improve the prediction performance of a model.

The aspects of network architecture and topology as well as preprocessing and postprocessing should be further investigated. Performance of a two-stage model may be improved further by appropriate syllable classification model and selection criterion of duration intervals. Speaker and language identity tasks are demonstrated with prosody models for a limited number of languages and speakers. One can extend this task with more number of speakers and languages to indicate the robustness. In the present work nonlinear (FFNN and SVM) models are used for prediction of prosody parameters. One can carry out the investigation with hybrid models (linear and nonlinear) and hybrid approaches (rule-based and statistical) for predicting the prosody.

For developing prosody models, manually labeled speech corpora is used in the present work. Labeling speech data is a tedious process. For example, labeling 5 hours of continuous speech data needs about 10,000 man-hours. In addition to this, the accuracy of labeling depends on human expert involved, since it is subjective. All these difficulties can be overcome by exploring ways to develop prosody models directly from unlabeled speech data.

In speech rate modification, exploiting the durational characteristics of various speech units with respect to their linguistic context and production constraints at different speaking rates may further improve the quality of speech after modification. For voice conversion application, both excitation source and vocal tract system parameters need to be modified according to specification of the target speech. In this book, methods are developed for incorporating the required formant structure and pitch contour. Modification of excitation source waveform (glottal waveform shape) is not addressed in the book. But it is necessary to control the shape of glottal pulse in each pitch period according to the desired specification. To carry out this, one needs to derive a relation between pitch, formant frequencies and glottal pulse specifications. After deriving the relationship, excitation waveform may also to be modified along with pitch and formant structure. This is likely to result in better quality of synthesized speech for voice conversion.

Appendix A
CODING SCHEME USED TO REPRESENT LINGUISTIC AND PRODUCTION CONSTRAINTS

A syllable is a combination of segments of consonants (C) and vowels (V). Each segment of the syllable is encoded separately. Codes for representing the segments (consonants or vowels) of the syllables are given in Table A.1. In this study, neural networks and support vector machines are used to model the prosody parameters from linguistic features. For developing these models, fixed length feature vectors are needed. Therefore we represented syllable with four segments. In this study, syllables with more than four segments are ignored. Syllables with less than four segments are represented by dummy segments. These dummy segments indicate the absence of segments, and are uniquely coded by a specific code '55'. The context of the syllable is represented by its adjacent (preceding and following) syllables. For the syllables at the word boundary (initial and final) have only one adjacent syllable is present as its context. In this case, the other syllable (missing) is represented by four dummy segments. These details can be observed in the following illustration.

K.S. Rao, *Predicting Prosody from Text for Text-to-Speech Synthesis*, SpringerBriefs
in Electrical and Computer Engineering, DOI 10.1007/978-1-4614-1338-7,
© Springer Science+Business Media New York 2012

Table A.1 Codes for representing the segments (consonants or vowels) of the syllables.

ai 58	ch 40	b	11
au 59	sh 41	c	12
a 60	Sh 42	d	13
i 61	kh 43	D	14
u 62	th 44	f	15
e 63	Th 45	g	16
o 64	ph 46	h	17
A 65	gh 47	j	18
I 66	dh 48	k	19
U 67	Dh 49	l	20
E 68	jh 50	L	21
O 69	bh 51	m	22
		n	23
		N	24
		p	25
		q	26
		r	27
		R	28
		s	29
		S	30
		t	31
		T	32
		v	33
		w	34
		x	35
		y	36
		z	37
		ñ	38
		Ñ	39
Absence of a segment: 55			

Illustration of feature extraction from the sequence of syllables in the text.

Text: *"pAkistAn ke pradhAn mantrI navAz sharIph"*
Table A.2 gives the following details:

1. Number of syllables in the given utterance.
2. Number of words in the utterance.
3. Position of the word in the utterance.
4. Position of the syllable in the utterance.
5. Position of the syllable in each word.

Table A.2 Syllable and word boundaries present in the utterance

	1st word			2nd word	3rd word		4th word		5th word		6th word	
Syllables	pA	kis	tAn	ke	pra	dhAn	man	trI	na	vAj	sha	rIph
Syllble position w.r.t word	1	2	3	1	1	2	1	2	1	2	1	2
Syllble position w.r.t phrase	1	2	3	4	5	6	7	8	9	10	11	12

Table A.3 illustrates the codes for 25 features for the syllable /pA/ in the utterance *"pAkistAn ke pradhAn mantrI navAz sharIph"*, which represent the positional, contextual and phonological information. They are as follows: 1, 3, 3, 1, 12, 12, 1, 6, 6, 55, 55, 55, 55, 19, 61, 29, 55, 25, 65, 55, 55, 1, 0, 2, 1.

Table A.3 Features for the syllable /pA/:

Linguistic and production constraints		Codes used for representing features
Syllable position position	word level	1 3 3
	phrase level	1 12 12
Word position	phrase level	1 6 6
Syllable context	previous syllable	55 55 55 55
	following syllable	19 61 29 55
Syllable identity		25 65 55 55
Syllable nucleus		1 0 2
Gender identity		1

Tables A.4, A.5 and A.6 illustrates the codes for 25 features for the syllables in the utterance "*pAkistAn ke pradhAn mantrI navAz sharIph*", which represent the positional, contextual and phonological information.

Table A.4 Positional features for the syllables present in the utterance "*pAkistAn ke pradhAn mantrI navAj sharIph*"

Syllables	Syllable position			Word position	
	Word	Phrase			
pA	1 3 3	1	12 12	1 6	6
kis	2 2 3	2	11 12	1 6	6
tAn	3 1 3	3	10 12	1 6	6
ke	1 1 1	4	9 12	2 5	6
pra	1 2 2	5	8 12	3 4	6
dhAn	2 1 2	6	7 12	3 4	6
man	1 2 2	7	6 12	4 3	6
trI	2 1 2	8	5 12	4 3	6
na	1 2 2	9	4 12	5 2	6
vAj	2 1 2	10	3 12	5 2	6
sha	1 2 2	11	2 12	6 1	6
rIph	2 1 2	12	1 12	6 1	6

Table A.5 Contextual features for the syllables present in the utterance "*pAkistAn ke pradhAn mantrI navAj sharIph*"

Syllables			Syllable identities			
Previous	Present	Following	Previous syllable		Following syllable	
–	pA	kis	55 55 55	55	19 61 29	55
pA	kis	tAn	25 65 55	55	31 65 23	55
kis	tAn	–	19 61 29	55	55 55 55	55
–	ke	–	55 55 55	55	55 55 55	55
–	pra	dhAn	55 55 55	55	48 65 23	55
pra	dhAn	–	25 27 60	55	55 55 55	55
–	man	trI	55 55 55	55	31 27 66	55
man	trI	–	22 60 23	55	55 55 55	55
–	na	vAj	55 55 55	55	33 65 18	55
na	vAj	–	23 60 55	55	55 55 55	55
–	sha	rIph	55 55 55	55	27 66 46	55
sha	rIph	–	41 60 55	55	55 55 55	55

Table A.6 Phonological features for the syllables present in the utterance *"pAkistAn ke pradhAn mantrI navAj sharIph"*

Syllable	Identity of syllable		Syllable nucleus			Gender
pA	25 65 55	55	1	0	2	1
kis	19 61 29	55	1	1	3	1
tAn	31 65 23	55	1	1	3	1
ke	19 63 55	55	1	0	2	1
pra	25 27 60	55	2	0	3	1
dhAn	48 65 23	55	1	1	3	1
man	22 60 23	55	1	1	3	1
trI	31 27 66	55	2	0	3	1
na	23 60 55	55	1	0	2	1
vAj	33 65 18	55	1	1	3	1
sha	41 60 55	55	1	0	2	1
rIph	27 66 46	55	1	1	3	1

References

1. S. Werner and E. Keller, "Prosodic aspects of speech," in *Fundamentals of Speech Synthesis and Speech Recognition: Basic Concepts, State of the Art, the Future Challenges* (E. Keller, ed.), pp. 23–40, Chichester: John Wiley, 1994.
2. K. S. Rao and B. Yegnanarayana, "Modeling syllable duration in Indian languages using neural networks," in *Proc. IEEE Int. Conf. Acoust., Speech, Signal Processing*, (Montreal, Quebec, Canada), pp. 313–316, May 2004.
3. K. S. Rao and B. Yegnanarayana, "Intonation modeling for Indian languages," in *Proc. Int. Conf. Spoken Language Processing*, (Jeju Island, Korea), pp. 733–736, Oct. 2004.
4. K. S. Rao and B. Yegnanarayana, "Prosodic manipulation using instants of significant excitation," in *Proc. IEEE Int. Conf. Multimedia and Expo*, (Baltimore, Maryland, USA), pp. 389–392, July 2003.
5. H. Mixdorff, *An integrated approach to modeling German prosody*. PhD thesis, Technical University, Dresden, Germany, July 2002.
6. D. H. Klatt, "Synthesis by rule of segmental durations in English sentences," in *Frontiers of Speech Communication Research* (B. Lindblom and S. Ohman, eds.), pp. 287–300, New York: Academic Press, 1979.
7. J. Allen, M. S. Hunnicut, and D. H. Klatt, *From Text to Speech: The MITalk system*. Cambridge: Cambridge University Press, 1987.
8. K. J. Kohler, "Zeitstrukturierung in der Sprachsynthese," *ITG-Tagung Digitalc Sprachverarbeitung*, no. 6, pp. 165–170, 1988.
9. K. Bartkova and C. Sorin, "A model of segmental duration for speech synthesis in French," *Speech Communication*, no. 6, pp. 245–260, 1987.
10. S. R. R. Kumar and B. Yegnanarayana, "Significance of durational knowledge for speech synthesis in Indian languages," in *Proc. IEEE Region 10 Conf. Convergent Technologies for the Asia-Pacific*, (Bombay, India), pp. 486–489, Nov. 1989.
11. S. R. R. Kumar, "Significance of durational knowledge for a text-to-speech system in an Indian language," Master's thesis, Dept. of Computer science and Engineering, Indian Institute of Technology Madras, Mar. 1990.
12. R. Sriram, S. R. R. Kumar, and B. Yegnanarayana, *A Text-to-Speech conversion system for Indian languages using parameter based approach*. Technical report no.12, Project VOIS, Dept. of CSE, IITM, May 1989.
13. K. K. Kumar, "Duration and intonation knowledge for text-to-speech conversion system for Telugu and Hindi," Master's thesis, Dept. of Computer science and Engineering, Indian Institute of Technology Madras, May 2002.
14. K. K. Kumar, K. S. Rao, and B. Yegnanarayana, "Duration knowledge for text-to-speech system for Telugu," in *Proc. Int. Conf. Knowledge Based Computer Systems*, (Mumbai, India), pp. 563–571, Dec. 2002.
15. S. H. Chen, W. H. Lai, and Y. R. Wang, "A new duration modeling approach for Mandarin speech," *IEEE Trans. Speech and Audio Processing*, vol. 11, pp. 308–320, July 2003.
16. J. P. H. V. Santen, "Segmental duration and speech timing," in *Computing prosody* (Sagisaka, Campbell, and Higuchi, eds.), pp. 225–249, Springer-Verlag, 1996.
17. J. P. H. V. Santen, "Assignment of segment duration in text-to-speech synthesis," *Computer Speech and Language*, vol. 8, pp. 95–128, Apr. 1994.
18. J. P. H. V. Santen, "Timing in text-to-speech systems," in *Proc. Eurospeech*, vol. 35, (Berlin, Germany), pp. 1397–1404, 1993.
19. J. P. H. V. Santen, "Analyzing n-way tables with sums-of-products models," *Journal of Mathematical Psychology*, vol. 37, pp. 327–371, 1993.
20. J. P. H. V. Santen, "Prosodic modeling in text-to-speech synthesis," in *Proc. Eurospeech*, (Rhodes, Greece), 1997.
21. J. P. H. V. Santen, C. Shih, and et. al., "Multi-lingual duration modeling," in *Proc. Eurospeech*, vol. 5, (Rhodes, Greece), pp. 2651–2654, 1997.

K.S. Rao, *Predicting Prosody from Text for Text-to-Speech Synthesis*, SpringerBriefs in Electrical and Computer Engineering, DOI 10.1007/978-1-4614-1338-7, © Springer Science+Business Media New York 2012

22. O. Goubanova and P. Taylor, "Using bayesian belief networks for model duration in text-to-speech systems," in *Proc. Int. Conf. Spoken Language Processing*, vol. 2, (Beijing, China), pp. 427–431, Oct. 2000.

23. O. Sayli, "Duration analysis and modeling for Turkish text-to-speech synthesis," Master's thesis, Dept. of Electrical and Electronics Engineering, Bogaziei University, 2002.

24. A. W. Black, P. Taylor, and R. Caley, "The Festival speech synthesis system." Manual and source code available at www.cstr.ed.ac.uk/projects/festival.html.

25. M. Riley, "Tree-based modeling of segmental durations," *Talking Machines: Theories, Models and Designs*, pp. 265–273, 1992.

26. S. Lee and Y.-H. Oh, "Tree-based modeling of prosodic phrasing and segmental duration for Korean TTS systems," *Speech Communication*, vol. 28, pp. 283–300, 1999.

27. A. Maghboulegh, "An empirical comparison of automatic decision tree and hand-configured linear models for vowel durations," in *Proc. of the Workshop in Computational Phonology in Speech Technology*, (Santa Cruz), 1996.

28. R. Batusek, "A duration model for Czech text-to-speech synthesis," in *Proceedings of TSD*, (Pilsen, Czech Republic), Sept. 2002.

29. H. Chung, "Segment duration in spoken Korean," in *Proc. Int. Conf. Spoken Language Processing*, (Denver, Colorado, USA), pp. 1105–1108, Sept. 2002.

30. N. S. Krishna and H. A. Murthy, "Duration modeling of Indian languages Hindi and Telugu," in *5th ISCA Speech Synthesis Workshop*, (Pittsburgh, USA), pp. 197–202, May 2004.

31. W. N. Campbell, "Analog i/o nets for syllable timing," *Speech Communication*, vol. 9, pp. 57–61, Feb. 1990.

32. W. N. Campbell, "Syllable based segment duration," in *Talking Machines: Theories, Models and Designs* (G. Bailly, C. Benoit, and T. R. Sawallis, eds.), pp. 211–224, Elsevier, 1992.

33. P. A. Barbosa and G. Bailly, "Characterization of rhythmic patterns for text-to-speech synthesis," *Speech Communication*, vol. 15, pp. 127–137, 1994.

34. P. A. Barbosa and G. Bailly, "Generation of pauses within the z-score model," in *Progress in Speech Synthesis*, pp. 365–381, Springer-Verlag, 1997.

35. R. Cordoba, J. A. Vallejo, J. M. Montero, J. Gutierrezarriola, M. A. Lopez, and J. M. Pardo, "Automatic modeling of duration in a Spanish text-to-speech system using neural networks," in *Proc. European Conf. Speech Communication and Technology*, (Budapest, Hungary), Sept. 1999.

36. Y. Hifny and M. Rashwan, "Duration modeling of Arabic text-to-speech synthesis," in *Proc. Int. Conf. Spoken Language Processing*, (Denver, Colorado, USA), pp. 1773–1776, Sept. 2002.

37. G. P. Sonntag, T. Portele, and B. Heuft, "Prosody generation with a neural network: Weighing the importance of input parameters," in *Proc. IEEE Int. Conf. Acoust., Speech, Signal Processing*, (Munich, Germany), pp. 931–934, Apr. 1997.

38. J. P. Teixeira and D. Freitas, "Segmental durations predicted with a neural network," in *Proc. European Conf. Speech Communication and Technology*, (Geneva, Switzerland), pp. 169–172, Sept. 2003.

39. R. E. Donovan, *Trainable speech synthesis*. PhD thesis, Cambridge University Engineering Department, Christ's college, Trumpington Street, Cambridge CB2 1PZ, England, June 1996.

40. A. Botinis, B. Granstrom, and B. Mobius, "Developments and paradigms in intonation research," *Speech Communication*, vol. 33, pp. 263–296, 2001.

41. P. A. Taylor, "Analysis and synthesis of intonation using the Tilt model," *Journal of Acoustic Society of America*, vol. 107, pp. 1697–1714, Mar. 2000.

42. P. A. Taylor, "The rise/fall/connection model of intonation," *Speech Communication*, vol. 15, no. 15, pp. 169–186, 1995.

43. J. B. Pierrehumbert, *The Phonology and Phonetics of English Intonation*. PhD thesis, MIT, MA, USA, 1980.

44. R. Sproat, ed., *Multilingual Text-to-Speech Synthesis: The Bell Labs Approach*. Kluwer Academic Publishers, 1998.

45. M. Jilka, G. Mohler, and G. Dogil, "Rules for generation of TOBI-based American English intonation," *Speech Communication*, vol. 28, pp. 83–108, 1999.

46. J. Buhmann, H. Vereecken, J. Fackrell, J. P. Martens, and B. V. Coile, "Data driven intonation modeling of 6 languages," in *Proc. Int. Conf. Spoken Language Processing*, vol. 3, (Beijing, China), pp. 179–183, Oct. 2000.

47. N. (Thorsen) Gronnum, "The groundworks of Danish intonation: An introduction." Museum Tusculanum Press, Copenhagen, 1992.

48. N. (Thorsen) Gronnum, "Superposition and subordination in intonation - a non-linear approach," in *Proceedings of the 13th International Congress - Phon. Sc. Stockholm*, (Stockholm), pp. 124–131, 1995.

49. E. Garding, "A generative model of intonation," in *Prosody: Models and Measuraments* (A. Cutler and D. R. Ladd, eds.), pp. 11–25, Berlin, Germany: Springer-Verlag, 1983.

50. H. Fujisaki, K. Hirose, P. Halle, and H. Lei, "A generative model for the prosody of connected speech in Japanese," in *Ann. Rep. Engineerng Research Institute 30*, pp. 75–80, 1971.

51. H. Fujisaki, "Dynamic characteristics of voice fundamental frequency in speech and singing," in *The production of speech* (P. F. MacNeilage, ed.), pp. 39–55, New York, USA: Springer-Verlag, 1983.

52. H. Fujisaki, "A note on the physiological and physical basis for the phrase and accent components in the voice fundamental frequency contour," in *Vocal Physiology: Voice Production, Mechanisms and Functions* (O. Fujimura, ed.), pp. 347–355, New York, USA: Raven Press, 1988.

53. H. Fujisaki, K. Hirose, and N. Takahashi, "Acoustic characteristics and the underlying rules of the intonation of the common Japanese used by radio and TV anouncers," in *Proc. IEEE Int. Conf. Acoust., Speech, Signal Processing*, pp. 2039–2042, 1986.

54. H. Fujisaki, S. Ohno, K. Nakamura, M. Guirao, and J. Gurlekian, "Analysis and synthesis of accent and intonation in standard Spanish," in *Proc. Int. Conf. Spoken Language Processing*, (Yokohama), pp. 355–358, 1994.

55. H. Fujisaki and S. Ohno, "Analysis and modeling of fundamental frequency contours of English utterances," in *Proceedings Eurospeech 95*, (Madrid), pp. 985–988, 1995.

56. H. Fujisaki, S. Ohno, and T. Yagi, "Analysis and modeling of fundamental frequency contours of Greek utterances," in *Proceedings Eurospeech 97*, (Rhodes, Greece), pp. 465–468, Sept. 1997.

57. H. Mixdorff and H. Fujisaki, "Analysis of voice fundamental frequency contours of German utterances using a quantitative model," in *Proc. Int. Conf. Spoken Language Processing*, vol. 4, (Yokohama), pp. 2231–2234, 1994.

58. P. Taylor and S. Isard, "A new model of intonation for use with speech synthesis and recognition," in *Proc. Int. Conf. Spoken Language Processing*, pp. 1287–1290, 1992.

59. J. t'Hart, R. Collier, and A. Cohen, *A perceptual study of intonation*. Cambridge: Cambridge University Press.

60. C. D'Alessandro and P. Mertens, "Automatic pitch contour stylisation using a model of tonal perception," *Computer Speech and Language*, vol. 9, pp. 257–288, 1995.

61. P. Mertens, *L'intonation du Franais: de la description linguistique a' la reconnaissance automatique*. PhD thesis, Katholieke Universiteit Leuven, Leuven, 1987.

62. J. Terken, "Synthesizing natural sounding intonation for Dutch: rules and perceptual evaluation," *Computer Speech and Language*, vol. 7, pp. 27–48, 1993.

63. J. R. de Pijper, "Modeling British English Intonation," 1983. Foris, Dordrecht.

64. L. M. H. Adriaens, *Ein Modell Deutscher Intonation*. PhD thesis, Technical University Eindhoven, Eindhoven, 1991.

65. C. Ode, "Russian intonation: A perceptual description," 1989. Rodopi, Amsterdam.

66. M. S. Scordilis and J. N. Gowdy, "Neural network based generation of fundamental frequency contours," in *Proc. IEEE Int. Conf. Acoust., Speech, Signal Processing*, vol. 1, (Glasgow, Scotland), pp. 219–222, May. 1989.

67. M. Vainio and T. Altosaar, "Modeling the microprosody of pitch and loudness for speech synthesis with neural networks," in *Proc. Int. Conf. Spoken Language Processing*, (Sidney, Australia), Dec. 1998.

68. M. Vainio, *Artificial neural network based prosody models for Finnish text-to-speech synthesis*. PhD thesis, Dept. of Phonetics, University of Helsinki, Finland, 2001.

69. S. H. Hwang and S. H. Chen, "Neural-network-based F0 text-to-speech synthesizer for Mandarin," *IEE Proc. Image Signal Processing*, vol. 141, pp. 384–390, Dec. 1994.

70. A. S. M. Kumar, S. Rajendran, and B. Yegnanarayana, "Intonation component of text-to-speech system for Hindi," *Computer Speech and Language*, vol. 7, pp. 283–301, 1993.

71. A. S. M. Kumar, *Intonation knowledge for speech systems for an Indian language*. PhD thesis, Dept. of Computer Science and Engineering, Indian Institute of Technology, Madras, Chennai, India, Jan. 1993.

72. T. F. Quatieri and R. J. McAulay, "Shape invariant time-scale and pitch modification of speech," *IEEE Trans. Signal Processing*, vol. 40, pp. 497–510, Mar. 1992.

73. E. Moulines and F. Charpentier, "Pitch-synchronous waveform processing techniques for text to speech synthesis using diphones," *Speech Communication*, vol. 9, pp. 453–467, Dec. 1990.

74. D. G. Childers, K. Wu, D. M. Hicks, and B. Yegnanarayana, "Voice conversion," *Speech Communication*, vol. 8, pp. 147–158, June 1989.

75. E. Moulines and J. Laroche, "Non-parametric techniques for pitch-scale and time-scale modification of speech," *Speech Communication*, vol. 16, pp. 175–205, Feb. 1995.

76. B. Yegnanarayana, S. Rajendran, V. R. Ramachandran, and A. S. M. Kumar, "Significance of knowledge sources for TTS system for Indian languages," *SADHANA Academy Proc. in Engineering Sciences*, vol. 19, pp. 147–169, Feb. 1994.

77. M. R. Portnoff, "Time-scale modification of speech based on short-time Fourier analysis," *IEEE Trans. Acoustics, Speech, and Signal Processing*, vol. 29, pp. 374–390, June. 1981.

78. M. R. Schroeder, J. L. Flanagan, and E. A. Lundry, "Bandwidth compression of speech by analytic-signal rooting," *Proc. IEEE*, vol. 55, pp. 396–401, Mar. 1967.

79. D. H. Klatt, "Review of text-to-speech conversion for English," *Journal of Acoustic Society of America*, vol. 82(3), pp. 737–793, Sep. 1987.

80. M. Narendranadh, H. A. Murthy, S. Rajendran, and B. Yegnanarayana, "Transformation of formants for voice conversion using artificial neural networks," *Speech Communication*, vol. 16, pp. 206–216, Feb. 1995.

81. E. P. Neuburg, "Simple pitch-dependent algorithm for high-quality speech rate changing," *Journal of Acoustic Society of America*, vol. 63, pp. 624–625, Feb. 1978.

82. E. B. George and M. J. T. Smith, "Speech Analysis/Synthesis and modification using an Analysis-by-Synthesis/Overlap-Add Sinusoidal model," *IEEE Trans. Speech and Audio Processing*, vol. 5, pp. 389–406, Sept. 1997.

83. R. Crochiere, "A weighted overlap-add method of short time Fourier analysis/synthesis," *IEEE Trans. Acoustics, Speech, and Signal Processing*, vol. 28, pp. 99–102, Feb. 1980.

84. S. Roucos and A. Wilgus, "High quality time-scale modification of speech," in *Proc. IEEE Int. Conf. Acoust., Speech, Signal Processing*, (Tampa, Florida, USA), pp. 493–496, Mar. 1985.

85. J. Laroche, Y. Stylianou, and E. Moulines, "HNS: Speech modification based on a harmonic + noise model," in *Proc. IEEE Int. Conf. Acoust., Speech, Signal Processing*, (Minneapolis, USA), pp. 550–553, Apr. 1993.

86. Y. Stylianou, "Applying the harmonic plus noise model in concatenative speech synthesis," *IEEE Trans. Speech and Audio Processing*, vol. 9, pp. 21–29, Jan. 2001.

87. H. Kawahara, "Speech representation and transformation using adaptive interpolation of weighted spectrum: Vocoder revisited," in *Proc. IEEE Int. Conf. Acoust., Speech, Signal Processing*, vol. 2, (Munich, Germany), pp. 1303–1306, 1997.

88. H. Kawahara, I. Masuda-Katsuse, and A. de Cheveigne, "Restructuring speech representations using a pitch-adaptive time-frequency smoothing and an instantaneous-frequency-based F0 extraction: Possible role of a repetitive structure in sounds," *Speech Communication*, vol. 27, pp. 187–207, 1999.

89. R. MuraliSankar, A. G. Ramakrishnan, and P. Prathibha, "Modification of pitch using DCT in source domain," *Speech Communication*, vol. 42, pp. 143–154, Jan. 2004.

90. R. MuraliSankar, A. G. Ramakrishnan, A. K. Rohitprasad, and M. Anoop, "DCT baced pitch modification," in *Proc. SPCOM 2001 6th Biennial Conference*, (Bangalore, India), pp. 114–117, July 2001.

91. W. Verhelst, "Overlap-add methods for time-scaling of speech," *Speech Communication*, vol. 30, pp. 207–221, 2000.

92. D. O'Brien and A. Monaghan, "Shape invariant time-scale modification of speech using a harmonic model," in *Proc. IEEE Int. Conf. Acoust., Speech, Signal Processing*, (Arizona, USA), 1999.

93. D. O'Brien and A. Monaghan, "Shape invariant pitch modification of speech using a harmonic model," in *Proc. Eurospeech*, (Budapest), 1999.

94. D. O'Brien and A. Monaghan, *Improvements in Speech Synthesis*, ch. Shape invariant pitch and time-scale modification of speech based on harmonic model. Chichester: John Wiley & Sons, 2001.

95. B. Yegnanarayana, C. d'Alessandro, and V. Darsinos, "An iterative algorithm for decomposition of speech signals into periodic and aperiodic components," *IEEE Trans. Speech and Audio Processing*, vol. 6, pp. 1–11, Jan. 1998.

96. S. Lemmetty, "Review of speech synthesis technology," Master's thesis, Dept. of Electrical and Communications Engineering, Helsinki University of Technology, Espoo, Finland, Mar. 1999.

97. R. Kortekaas and A. Kohlrausch, "Psychoacoustical evaluation of the Pitch Synchronous Overlap-and-Add speech waveform manipulation technique using single formant stimuli," *Journal of Acoustic Society of America*, vol. 101, no. 4, pp. 2202–2213, 1997.

98. H. Valbret, E. Moulines, and J. P. Tubach, "Voice transformation using PSOLA techniques," *Speech Communication*, vol. 11, pp. 175–187, 1992.

99. Y. Jiang and P. Murphy, "Production based pitch modification of voiced speech," in *Proc. Int. Conf. Spoken Language Processing*, (Denver, Colorado, USA), pp. 2073–2076, Sept. 2002.

100. S. Haykin, *Neural Networks: A Comprehensive Foundation*. New Delhi, India: Pearson Education Aisa, Inc., 1999.

101. B. Yegnanarayana, *Artificial Neural Networks*. New Delhi, India: Printice-Hall, 1999.

102. V. N. Vapnik, *Statistical Learning Theory*. New York: Wiley, 2001.

103. A. Smola and B. Scholkopf, *A Tutorial on Support Vector Regression*. Technical report Neuro COLT NC-TR-98-030, 1998.

104. X. Huang, A. Acero, and H. W. Hon, *Spoken Language Processing*. New York, NJ, USA: Prentice-Hall, Inc., 2001.

105. D. H. Klatt, "Linguistic uses of segmental duration in English: Acoustic and perceptual evidence," *Journal of Acoustic Society of America*, vol. 59, pp. 1209–1221, 1976.

106. A. W. F. Huggins, "On the perception of temporal phenomena in speech," *Journal of Acoustic Society of America*, vol. 4, pp. 1279–1290, 1972.

107. D. K. Oller, "The effect of position in utterance on speech segment duration in English," *Journal of Acoustic Society of America*, vol. 54, pp. 1247–1253, 1973.

108. T. H. Crystal and A. S. House, "Characterization and modeling of speech segment durations," in *Proc. IEEE Int. Conf. Acoust., Speech, Signal Processing*, pp. 2791–2794, 1986.

109. T. H. Crystal and A. S. House, "The duration of American English vowels: an overview," *Journal of Phonetics*, vol. 16, pp. 263–284, 1988.

110. T. H. Crystal and A. S. House, "The duration of American English stop consonants: An overview," *Journal of Phonetics*, vol. 16, pp. 285–294, 1988.

111. K. N. Reddy, "The duration of Telugu speech sounds: an acoustic study," *Special issue of Journal of IETE on Speech processing*, pp. 57–63, 1988.

112. S. R. Savithri, "Duration analysis of Kannada vowels," *Journal of Acoustical Society of India*, vol. 4, pp. 34–40, 1986.

113. K. S. Rao, S. V. Gangashetty, and B. Yegnanarayana, "Duration analysis for Telugu language," in *Int. Conf. Natural Language Processing*, (Mysore, India), pp. 152–158, Dec. 2003.

114. N. Umeda, "Linguistic rules for text-to-speech synthesis," *Proc. IEEE*, vol. 4, pp. 443–451, 1976.

115. A. Chopde, "Itrans Indian language transliteration package version 5.2 source." http://www.aczone.con/itrans/.

116. A. N. Khan, S. V. Gangashetty, and B. Yegnanarayana, "Syllabic properties of three Indian languages: Implications for speech recognition and language identification," in *Int. Conf. Natural Language Processing*, (Mysore, India), pp. 125–134, Dec. 2003.

117. O. Fujimura, "Syllable as a unit of speech recognition," *IEEE Trans. Acoustics, Speech, and Signal Processing*, vol. 23, pp. 82–87, Feb. 1975.

118. K. S. Rao and S. G. Koolagudi, "Selection of suitable features for modeling the durations of syllables," *Journal of Software Engineering and Applications*, vol. 3, Dec. 2010.

119. M. Riedi, "A neural network based model of segmental duration for speech synthesis," in *Proc. European Conf. Speech Communication and Technology*, (Madrid), pp. 599–602, Sept. 1995.

120. W. N. Campbell, "Predicting segmental durations for accommodation within a syllable-level timing framework," in *Proc. European Conf. Speech Communication and Technology*, vol. 2, (Berlin, Germany), pp. 1081–1084, Sept. 1993.

121. S. Rajendran, K. S. Rao, B. Yegnanarayana, and K. N. Reddy, "Syllable duration in broadcast news in Telugu: A preliminary study," in *National Conf. on Language Technology Tools: Implementation of Telugu/Urdu*, (Hyderabad, India), Oct. 2003.

122. K. S. Rao, S. V. Gangashetty, and B. Yegnanarayana, "Duration analysis for Telugu language," in *Int. Conf. on Natural Language Processing (ICON)*, (Mysore, India), Dec. 2003.

123. S. Lee, K. Hirose, and N. Minematsu, "Incoporation of prosodic modules for large vocabulary continuous speech recognition," in *Proc. ISCA Workshop on Prosody in Speech recognition and understanding*, pp. 97–101, 2001.

124. K. Ivano, T. Seki, and S. Furui, "Noise robust speech recognition using F0 contour extract by Hough transform," in *Proc. Int. Conf. Spoken Language Processing*, pp. 941–944, 2002.

125. L. Mary and B. Yegnanarayana, "Prosodic features for speaker verification," in *Proc. Int. Conf. Spoken Language Processing*, (Pittsburgh, PA, USA), pp. 917–920, Sep. 2006.

126. L. Mary, *Multi level implicit features for language and speaker recognition*. PhD thesis, Dept. of Computer Science and Engineering, Indian Institute of Technology Madras, Chennai, India, June 2006.

127. L. Mary and B. Yegnanarayana, "Consonant-vowel based features for language identification," in *Int. Conf. Natural Language Processing*, (Kanpur, India), pp. 103–106, Dec. 2005.

128. L. Mary, K. S. Rao, and B. Yegnanarayana, "Neural network classifiers for language identification using phonotactic and prosodic features," in *Proc. Int. Conf. Intelligent Sensing and Information Processing (ICISIP)*, (Chennai, India), pp. 404–408, Jan. 2005.

129. K. K. Kumar, K. S. Rao, and B. Yegnanarayana, "Duration knowledge for text-to-speech system for telugu," in *Int. Conf. Knowledge based computer systems (KBCS)*, (Mumbai, India), Dec. 2002.

130. C. J. C. Burges, "A tutorial on support vector machines for pattern recognition," *Data Mining and Knowledge Discovery*, vol. 2, no. 2, pp. 121–167, 1998.

131. T. B. Trafalis and H. Lnce, "Support vector machine for regression and applications to financial forecasting," in *Int. Joint Conf. Neural Networks*, pp. 348–353, June 2000.

132. J. R. Bellegarda, K. E. A. Silverman, K. Lenzo, and V. Anderson, "Statistical prosodic modeling: From corpus design to parameter estimation," *IEEE Trans. Speech and Audio Processing*, vol. 9, pp. 52–66, Jan. 2001.

133. J. R. Bellegarda and K. E. A. Silverman, "Improved duration modeling of English phonemes using a root sinusoidal transformation," in *Proc. Int. Conf. Spoken Language Processing*, pp. 21–24, Dec. 1998.

134. K. E. A. Silverman and J. R. Bellegarda, "Using a sigmoid transformation for improved modeling of phoneme duration," in *Proc. IEEE Int. Conf. Acoust., Speech, Signal Processing*, (Phoenix, AZ, USA), pp. 385–388, Mar. 1999.

135. B. Siebenhaar, B. Zellner-Keller, and E. Keller, "Phonetic and timing considerations in a Swiss high German TTS system," in *Improvements in Speech Synthesis* (E. Keller, G. Bailly, A. Monaghan, J. Terken, and M. Huckvale, eds.), Chichester: John Wiley, 2001.

136. C. S. Gupta, S. R. M. Prasanna, and B. Yegnanarayana, "Autoassociative neural network models for online speaker verification using source features from vowels," in *Int. Joint Conf. Neural Networks*, (Honululu, Hawii, USA), May 2002.

137. B. Yegnanarayana and S. P. Kishore, "AANN an alternative to GMM for pattern recognition," *Neural Networks*, vol. 15, pp. 459–469, Apr. 2002.
138. K. S. Rao and B. Yegnanarayana, "Modeling syllable duration in Indian languages using neural networks," in *Proc. IEEE Int. Conf. Acoust., Speech, Signal Processing*, (Montreal, Quebec, Canada), May 2004.
139. K. S. Rao and B. Yegnanarayana, "Modeling syllable duration in indian languages using support vector machines," in *2nd Int. Conf. Intelligent Sensing and Information Processing (ICISIP-2005)*, (Chennai, India), Jan. 2005.
140. K. S. Rao and B. Yegnanarayana, "Modeling durations of syllables using neural networks," *Computer Speech and Language*, vol. 21, pp. 282–295, Apr. 2007.
141. K. S. Rao, "Modeling supra-segmental features of syllables using neural networks," in *Speech, Audio, Image and Biomedical Signal Processing using Neural Networks* (P. B. Prasad and S. R. M. Prasanna, eds.), pp. 71–95, Springer, 2008.
142. K. S. Rao and B. Yegnanarayana, "Impact of constraints on prosody modeling for Indian languiages," in *3rd International Conference on Natural Language Processing (ICON-2004)*, (Hyderabad, India), pp. 225–236, Dec. 2004.
143. K. S. Rao and B. Yegnanarayana, "Two-stage duration model for indian languages using neural networks," in *Lecture Notes in Computer Science, Neural Information Processing (Springer)*, pp. 1179–1185, 2004.
144. K. S. Rao, "Application of prosody models for developing speech systems in Indian languages," *International Journal of Speech Technology*, vol. 14, pp. 19–23, March 2011.
145. S. R. M. Prasanna and B. Yegnanarayana, "Extraction of pitch in adverse conditions," in *Proc. IEEE Int. Conf. Acoust., Speech, Signal Processing*, (Montreal, Canada), May 2004.
146. K. S. Rao and B. Yegnanarayana, "Intonation modeling for indian languages," *Computer Speech and Language*, vol. 23, pp. 240–256, Apr. 2009.
147. K. S. Rao and B. Yegnanarayana, "Intonation modeling for indian languages," in *8th Int. Conf. on Spoken Language Processing (Interspeech-2004)*, (Jeju Island, Korea), pp. 733–736, Oct. 2004.
148. L. Mary, K. S. Rao, S. V. Gangashetty, and B. Yegnanarayana, "Neural network models for capturing duration and intonation knowledge for language and speaker identification," in *8th Int. Conf. on Cognitive and Neural systems (ICCNS)*, (Boston, MA, USA), May 2004.
149. L. Mary, K. S. Rao, and B. Yegnanarayana, "Neural network classifiers for language identification using syntactic and prosodic features," in *2nd Int. Conf. Intelligent Sensing and Information Processing (ICISIP-2005)*, (Chennai, India), Jan. 2005.
150. S. G. Koolagudi and K. S. Rao, "Neural network models for capturing prosodic knowledge for emotion recognition," in *12th Int. Conf. on Cognitive and Neural systems (ICCNS)*, (Boston, MA, USA), May 2008.
151. P. S. Murthy and B. Yegnanarayana, "Robustness of group-delay-based method for extraction of significant excitation from speech signals," *IEEE Trans. Speech and Audio Processing*, vol. 7, pp. 609–619, Nov. 1999.
152. R. Smits and B. Yegnanarayana, "Determination of instants of significant excitation in speech using group delay function," *IEEE Trans. Speech and Audio Processing*, vol. 3, pp. 325–333, Sept. 1995.
153. J. Makhoul, "Linear prediction: A tutorial review," *Proc. IEEE*, vol. 63, pp. 561–580, Apr. 1975.
154. A. V. Oppenheim, R. W. Schafer, and J. R. Buck, *Discrete-time signal processing*. Upper Saddle River, NJ.: Prentice-Hall, 1999.
155. W. M. Fisher, G. R. Doddington, and K. M. Goudie-Marshall, "The DARPA speech recognition database: Specifications and status," in *Proc. DARPA Workshop on speech recognition*, pp. 93–99, Feb. 1986.
156. J. R. Deller, J. G. Proakis, and J. H. L. Hansen, *Discrete-time processing of speech signals*. New York, USA: Macmilan Publishing Company, 1993.
157. R. V. Hogg and J. Ledolter, *Engineering Statistics*. 866 Third Avenue, New York, USA: Macmillan Publishing Company, 1987.
158. K. S. Rao and B. Yegnanarayana, "Prosody modification using instants of significant excitation," *IEEE Trans. Speech and Audio Processing*, vol. 14, pp. 972–980, May 2006.

159. K. S. Rao and B. Yegnanarayana, "Prosodic manipulation using instants of significant excitation," in *Proc. IEEE Int. Conf. Acoust., Speech, Signal Processing*, April 2003.

160. K. S. Rao and B. Yegnanarayana, "Prosodic manipulation using instants of significant excitation," in *IEEE Int. Conf. Multimedia and Expo*, (Baltimore, Maryland, USA), July 2003.

161. T. V. Ananthapadmanabha and B. Yegnanarayana, "Epoch extraction from linear prediction residual for identification of closed glottis interval," *IEEE Trans. Acoustics, Speech, and Signal Processing*, vol. 27, pp. 309–319, Aug. 1979.

162. A. V. Oppenheim and R. W. Schafer, *Digital Signal Processing*. Englewood Cliffs, New Jersey, USA: Prentice Hall, 1975.

163. B. Yegnanarayana, S. R. M. Prasanna, and K. S. Rao, "Speech enhancement using excitation source information," in *Proc. IEEE Int. Conf. Acoust., Speech, Signal Processing*, vol. 1, (Orlando, Florida, USA), pp. 541–544, May 2002.

164. D. Gabor, "Theory of communication," *J. IEE*, vol. 93, no. 2, pp. 429–457, 1946.

165. N. S. Krishna, H. A. Murthy, and T. A. Gonsalves, "Text-to-speech (tts) in indian languages," in *Int. Conf. Natural Language Processing*, 2002.

166. S. Srikanth, S. R. R. Kumar, R. Sundar, and B. Yegnanarayana, *A text-to-speech conversion system for Indian languages based on waveform concatenation model*. Technical report no.11, Project VOIS, Dept. of Computer Science and Engineering, Indian Institute of Technology Madras, Mar. 1989.

167. B. Zellner, "Fast and slow speech rate: A characterization for French," in *Proc. Int. Conf. Spoken Language Processing*, (Sydney, Australia.), pp. 542–545, Dec. 1998.

168. S. R. M. Prasanna and J. M. Zachariah, "Detection of vowel onset point in speech," in *Proc. IEEE Int. Conf. Acoust., Speech, Signal Processing*, (Orlando, Florida, USA), May 2002.

169. S. V. Gangashetty, C. C. Sekhar, and B. Yegnanarayana, "Extraction of fixed dimension patterns from varying duration segments of consonant-vowel utterances," in *Proc. IEEE Int. Conf. Intelligent Sensing and Information Processing*, (Chennai, India), pp. 159–164, Jan. 2004.

170. *Database for Indian languages*. Speech and Vision lab, Indian Institute of Technology Madras, India, 2001.

171. H. A. Murthy and B. Yegnanarayana, "Formant extraction from group delay function," *Speech Communication*, vol. 10, pp. 209–221, Mar. 1991.

172. K. S. Rao, S. R. M. Prasanna, and B. Yegnanarayana, "Determination of instants of significant excitation in speech using hilbert envelope and group delay function," *IEEE Signal Processing Letters*, vol. 14, pp. 762–765, Oct. 2007.

173. K. S. Rao, "Real time prosody modification," *Journal of Signal and Information Processing*, Nov. 2010.

174. K. S. Rao and B. Yegnanarayana, "Neural network models for text-to-speech synthesis," in *5th International Conference on Knowledge Based Computer Systems (KBCS-2004)*, (Hyderabad, India), pp. 520–530, Dec. 2004.

175. K. S. Rao and B. Yegnanarayana, "Duration modification using glottal closure instants and vowel onset points," *Speech Communication*, vol. 51, pp. 1263–1269, Dec. 2009.

176. K. S. Rao and B. Yegnanarayana, "Voice conversion by prosody and vocal tract modification," in *9th Int. Conf. Information Technology*, (Bhubaneswar, Orissa, India), Dec 2006.

177. K. S. Rao, "Voice conversion by mapping the speaker-specific features using pitch synchronous approach," *Computer Speech and Language*, vol. 24, pp. 474–494, July 2010.